REACHING
DOWN THE
RABBIT HOLE

REACHING DOWN THE RABBIT HOLE

A Renowned Neurologist
Explains the Mystery and Drama
of Brain Disease

Dr. Allan H. Ropper and
Brian David Burrell

St. Martin's Press
New York

www.stmartins.com

Designed by Kathryn Parise

LIBRARY OF CONGRESS CATALOGING-IN-PUBLICATION DATA

Ropper, Allan H.
  Reaching down the rabbit hole : a renowned neurologist explains the mystery and drama of brain disease / Dr. Allan H. Ropper and Brian David Burrell.
      p. cm.
    ISBN 978-1-250-03498-4 (hardcover)
    ISBN 978-1-250-03499-1 (e-book)
    1. Ropper, Allan H.   2. Neurology—Anecdotes.   3. Brain—Diseases—Anecdotes.   4. Neurologists—Massachusetts—Boston—Biography.   I. Burrell, Brian, 1955–   II. Title.
    RC351.R66 2014
    616.8—dc23

                                                              2014017011

St. Martin's Press books may be purchased for educational, business, or promotional use. For information on bulk purchases, please contact Macmillan Corporate and Premium Sales Department at 1-800-221-7945, extension 5442, or write specialmarkets@macmillan.com.

First Edition: September 2014

10 9 8 7 6 5 4 3 2 1

CONTENTS

REACHING DOWN THE RABBIT HOLE

INTRODUCTION:
The Queen

What neurologists do

"Hello, I'm Dr. Allan Ropper. How are you?"

"That's a grammatical question. Plus and minus."

"Is your mind clear?"

"I guess so. There are a lot of unexplained issues around me, but . . . I'm communicating appropriately."

His name is Dr. Vandermeer. He is in his mideighties, and he is a genuine Boston Brahmin. I know his type well. Over the past fifty years he has built a national reputation as a top-drawer researcher and a caring physician, universally admired, all the while taking somewhat heedless care of his own body. He is a man of arts and sciences, but also a man of tastes and habits acquired from his father and grandfather, which is to say, he is a nineteenth-century Yankee living a twentieth-century life in the twenty-first century, and is only vaguely aware that he can no longer manage in the ten-room house he has occupied for the last fifty-two years. When he retired in his seventies, he settled upon a daily routine that failed to anticipate his declining faculties. He is as unwilling to accept this fact as he is to acknowledge the unruliness of his eyebrows or his surplus of nose and ear hair, a clinical indifference that is not unusual in aging doctors.

"Do you know where you are?"

"At Brigham and Women's Hospital."

"The date?"

"The date? No, I couldn't give you that."

"Year?"

"Again, it's such a confusing sequence of events that it has confounded me to where my orientation isn't what it should be."

"No problem. Are you in any pain?"

"No."

Doc Vandermeer was brought here over his own objections after his wife found him sitting on the toilet some seven hours after he had ventured from bedroom to bathroom. He had spent the night there.

"Have you had any hallucinations?"

"I don't think so, but then people generally don't."

"Touché. Have you had any convulsions?"

"No."

"And you know about this meningioma in your right temporal lobe? You have a meningioma there about the size of a lemon. Were you aware of that?"

"I've had two growth issues that are pertinent. One is in the pancreas and the other is there."

"But you seem a little off, cognitively, and our struggle is, could it be the meningioma that's doing it?"

"As they say, that's your problem."

He's right. It *is* my problem. I am his neurologist. It is my job to parse his convoluted responses, fit them into the clinical picture, prioritize his issues, and come up with a plan that lets him live the life that he wants to live, to the extent that is possible. In his peculiar way, he has indicated that he is oriented to his location, but not oriented to the time or to his situation. His highly formalized locution, entirely characteristic of his tribe, may seem quaint, but it is exaggerated just enough to suggest that he is compensating for a language deficit that he is aware of but seems indifferent to. He knows that he has a benign tumor in his pancreas that poses less of a threat than the meningioma in his brain. The brain tumor will not kill him anytime soon, but it will continue to diminish his powers of thought.

"We'll find out what's what and let you know," I say. "Great to meet you."

It will take us some time to sort it all out, but this is a start. Hannah, my senior resident, gives me a nod, signaling that it's time to move on to the next patient.

"Hello, there. I'm Dr. Ropper. I'm one of the neurologists. And you've met Hannah Ross, our chief resident. Do you mind if we visit with you?"

His name is Gary, and he is one unhappy customer. He is thirty-two years old and has spent more time in hospitals than he cares to recall.

"How are you doing?" I ask.

"I can function," Gary replies listlessly, "but not to design specifications."

"Interesting phrase. Where did you get that from?"

"Gene Roddenberry."

It fits. Gary is a computer programmer who has so immersed himself in *Star Trek* that he knows the Starship Design Specifications by heart. He has the body of a Buddha, the eyes of a koala, and an air of resignation that is probably his baseline. He might as well be wearing a T-shirt emblazoned with the motto: IT IS WHAT IT IS. He has had epilepsy since childhood, culminating in the surgical removal of part of his brain when he was seventeen years old. A U-shaped scar stands out prominently on the left side of his close-cropped scalp. He has enough experience with his condition to know when he needs to go to the hospital, although he would rather not go. When he does, he most definitely does not want to be admitted, but Hannah has admitted him anyway, and there is a lingering animus on his part.

"I understand you had several events yesterday," Hannah says. "A lot of seizures?"

"Well, not many for me. What's considered many?"

"It all depends on how many is many for you. I was under the impression that you did have many for you . . . in a *row*. Is that not true?"

"No. Even when I have clusters it's probably four or five."

"Is that what you had yesterday? Four or five?"

"Well, that's not what brought me to the hospital, if that's what you're getting at."

Gary is in a mood, so I'm happy to let Hannah carry the ball. This is her ward, these are her patients, I am the attending physician, and while I took the lead with Dr. Vandermeer in deference to his stature, I now take up a position at the foot of the bed and watch.

"That *is* what I'm getting at," Hannah continues in a sing-song tone as if trying to engage a curious child. "So what brought you to the hospital today?"

A pause. Gary seems bothered less by Hannah's tone than by her question. "It was late in the afternoon," he replies with mock sarcasm. "I had a really bad headache and I was feeling very tired. I was clumsy and I couldn't stand up straight and my hands were vibrating."

"And that's not what usually happens with your seizures?"

"Well, I don't get so much warning with the seizures, although some days I can tell if I'm walking up to the edge of a seizure, but it doesn't always lead to a seizure. Sometimes it leads to a lot of seizures, sometimes nothing."

"How many seizures do you usually have in a given day?"

"There is no usual. I can go as long as five days without any seizures at all."

Hannah's directed line of inquiry seems to have snared his attention. He is clearly not used to having anyone take such a keen interest in his troubles.

"And what's your usual when you do have them?" she asks.

"My record is eighty-four seizures in a day."

"Is that typical?"

"There is no rhyme or reason to it. Some days I have none, some days I have a lot."

"But say in the course of a month."

"You keep *wanting* me to do that. You keep *wanting* me to put some pattern to it. Believe me, I wish I *could* put a pattern to it. Would you like to see my seizure log?"

"That won't be necessary."

I'm staying out of it. Hannah is firm yet restrained and unflustered; Gary is exasperated to the point of annoyance, almost anger, without the emotional edge to attain real anger. He acts besieged, yet lacks the reserves of panic that fuel true paranoia. It's as though he has programmed himself to be confrontational, but the confrontational part of his brain is so compromised that it can't run the code.

Hannah presses on. "What I really want to understand is whether you ever have a month without seizures."

"A month? No. I'm lucky if I go a week."

"And then when they come you can have one or you can have a ton?"

"Yes."

"Any missed doses of your meds?"

"Everyone wants to blame the epileptic. 'You wouldn't have seizures if you didn't miss your doses.' Is that it?"

She doesn't bite. "No. You'd probably have seizures anyway."

"That's true. I don't miss doses, and I have seizures."

We have the picture now. Fortunately, Gary does not. If he did, he would know that Hannah suspects, as do I, that some of his seizures are fake. It is time for us to get going again.

"It was really great to meet you," I say. "I wish the rest of our patients had your outlook."

"You mean felt that there's nothing you guys can do?"

"No, I mean had your level of self-awareness."

Gary has what was once called a temporal-lobe seizure personality. He is intelligent, mildly paranoid, has good word usage, and most of all is extremely literal. He is wrong that there's nothing we can do, although this has long been a knock on neurology: that we can usually tell what's going on in someone's brain, but that we can't do much about it. I won't bother to disabuse him of the idea because it would take a book to do so. This book, for example.

Gary is right about one thing. His mind is not working to design specifications, although in that regard he is hardly unique. There are twenty-nine other people on the ward right now who could say the same thing.

They call this place the Brigham, short for Brigham and Women's Hospital, an amalgam of the names of a few historic hospitals that merged decades ago—the Peter Bent Brigham, the Robert Breck Brigham, and the Boston Hospital for Women—and it spans the length of a city block that abuts the Harvard Medical School campus. This massive enclave of teaching hospitals, which also includes Beth Israel Deaconess Hospital, the Dana Farber Cancer Institute, Boston Children's Hospital, and the Joslin Diabetes Center, serves as a training ground for Harvard's medical students. It is a city within a city, or at least a city on the very edge of a city: a district of glass, stone, and steel towers that occupies the eastern bank of a lazy, muddy waterway that separates Boston from the suburb of Brookline. At clinics and community hospitals throughout New England, whenever a doctor looks over a patient chart, sighs, and says, "Send it to Boston," there is a good chance that the patient will end up here.

On any weekday morning the inbound traffic is daunting, like a rising tide in a teeming estuary, bearing with it an influx of patients, visitors, and family members, not to mention staff, that will be carried away with the ebb at day's end, leaving behind isolated pools and puddles of low-level activity.

This is one of those pools—the neurology inpatient ward. It occupies the tenth floor of the hospital tower. Built in the 1980s, the tower has a cross section in the shape of a four-leafed clover, each leaf being a pod, each pod consisting of a dozen fanned-out rooms, each room visible and easily accessible from a semicircular nurses' station, a novel idea that now seems somewhat dated. The patients here, like Gary and Doc Vandermeer, are in tough shape, and they spend much of their morning waiting for our visit, and the rest of the day wishing they were someplace else, even though this is the best place they could be. It is a place where the strangest and most challenging cases are sent to be sorted out, cases whose complexity would tax the resources of smaller hospitals.

I am a clinical neurologist and a professor of neurology. Most people have little idea what that means, but among other things it means that I am an authority on what the brain does right and does wrong: language, sensation, and emotion; walking, falling, weakness, tremors, and coordination; memory, mental incapacity, delays in development; anxiety, pain, stress, even death. The practice of my craft, the clinical part of it, is the systematic, logical, deductive method that was in the past applicable to all branches of medicine, but now resides mainly in neurology. The paradoxical part of it, the unique challenge, is that my primary sources of information—my patients' brains—are quite often altered, sometimes bizarrely, as a result of disease. This creates an incredible and self-referential conundrum. How do you begin to understand a sick brain? The only viable answer, as this book will show, is that you do it by engaging the person inside, and you do it on a case-by-case basis.

"He hasn't felt emotion for at least a month."

This is the girlfriend speaking. She is in her midthirties, as is her boyfriend, the man without emotion. They are not well-to-do, but they get by. She speaks with a strong North Shore accent, and despite her downscale style of dress, she has above-average verbal skills: she listens carefully, processes information quickly, and answers helpfully.

They have lived together for five years. They have a child. They are resigned to their fate.

"What do you mean by that?" I ask her.

"No sadness over what's happening. No anger anymore. When he was first diagnosed he had a lot of anger issues. But recently he's just been happy, not out of control."

"Happy, or just nothing?"

"I would say content," she replies. "He's easygoing. The TV used to make him laugh. I saw happiness then. Not now."

"When he expresses an emotion . . . ?"

"He may smile, nothing else."

"In speaking to you, speaking to each other?"

"No. He really hasn't initiated a conversation in the last month."

"That's typical for a butterfly glioma infiltrating the frontal lobes," I say, referring to his spreading, malignant brain tumor. "So we're really seeing him in good shape now. Is that true?"

"Yeah, he's better right now, on steroids. This is way better than he's been in a long time now."

I turn to him. "Can you tell us a joke, Dennis?"

"Hopefully not," he replies with little affect. Dennis seems indifferent even when he grins. His head is shaved, and his cranial stubble reveals the dark shadow of a receding hairline. He has a tattoo on his left forearm, in Gothic text, that reads: 𝕾𝕴𝕮𝕶.

"There's enough rattling around in there so that your thinking seems really good."

"Yeah."

"What's missing? What's not right in your head?"

Dennis pauses, inhales deeply through his nostrils, thinks about it, and lets it out slowly, but says nothing.

"You have trouble expressing yourself in language, or expressing yourself in emotion?"

He takes his time. "I think it's in emotion."

To the girlfriend again: "I understood from the residents that there's a conversation going on about hospice. How did you get to that? Was he becoming unmanageable at home?"

"Over the last few weeks, yes, he was."

"On account of what?"

"Behavior. He wouldn't listen to me. He kept getting up and opening the fridge, wanting to take his medicine. Then just being a little bit agitated. But as far as stopping treatment, we talked to the oncologist last night. The chemo affected him very badly. The doctor said that there aren't any drugs left that won't make his life really terrible. So we're talking about quality of life at this point, because the tumor is so large, and it's inoperable."

I turn to Dennis: "Are you engaged in these conversations?"

"Not that one, but we did talk about treatment a bunch yesterday with Dr. Nadgir."

"Before," she says, "he wasn't on a high dose of steroids, he wasn't clear at all. But now, on the steroids, this is a complete turnaround."

"Are your plans staying the same?" I ask her.

"I'm thinking maybe we can bring him home and have hospice come in."

"Can you behave yourself if you go home?" I say to Dennis.

He grins and shakes his head in a "no" gesture.

"No? So this is some combination of the tumor and his character. I mean, you *are* a character, Dennis, right?"

His grin springs from memory more than feeling. He knows that it's only a matter of months. The irony is that the location of the tumor has neutralized the part of his brain that would allow him to care. So he *was* a character, but not so much anymore.

What Dennis, his girlfriend, and everybody else on the ward need, more than anything else, is to tell us their stories. Many of them have driven for an hour, two, even three hours to the Hub of the Universe (as Boston styles itself), and they want to be heard. What they hope, what they expect, what they deserve, is that we take the time to listen, because the act of listening is therapeutic in itself. When we do it right, we learn details that make us better doctors for the next patient. The residents may not get this yet. They are focused on diagnosis and treatment, on technology, on scales, titers, doses, ratios, elevations, and deficiencies. All well and good, I tell them, but don't forget to listen.

Hannah and the team are huddled around a computer monitor in a cubbyhole near the nurses' station on the tenth floor. They have made it through medical school and earned their MDs, chosen to go into neurology, and this is now finishing school for them, a prestigious appointment at a top-tier hospital. My role is to look over their shoulders, provide them with someone to model, give them a hard time (or, pedagogically speaking, challenge their assumptions).

As they play at the computer, a patient sits twenty feet away, just beyond a glass door and curtains. He was admitted three hours ago with the sudden onset of a speech difficulty and a complete personality change. The team hasn't visited him yet. Instead, they are transfixed by the images of his brain up on the screen, like passengers glued to the in-flight movie as they fly over the Grand Canyon at sunrise. It reminds me of an old joke: Two grandmothers meet on the boardwalk, one pushing a baby carriage. "That's a beautiful granddaughter," the friend says, and the other replies, "This is nothing, you should see her pictures."

"Anything?" I ask.

"I don't think it's a stroke," Hannah says. "Maybe there's a low-grade glioma in the right frontal lobe, a little bit of hypodensity."

"Would you put a dime on anything now?" I ask her.

"Tumor."

"What do you think?" I ask Rakesh, a second-year resident.

"I think it could be a combination of tumor and postictal changes," he replies. "How about you?"

"Me?" I say. "I'll be glad to hold the money for you." What I want is for them to step away from the monitor and into the room, sit by the bed, talk to the patient, and examine the man instead of his pixels.

"Here's a small side bet," I add. "I don't think the enhanced MRI is going to help us. Look, you've got the whole package here—the scans and the tests—and you're still not sure what's going on. That's why neurology is the queen of the medical specialties."

"Not the king?"

"No, it's the queen because the queen is elegant, and it is the last place in medicine where your personal synthetic intellectual effort is value added, and despite all these gee-whiz gizmos, there really are no tests. You're going to have to sort this one out at the bedside."

Hannah has heard this speech before, and has a pretty good idea what it means. It means that if a patient comes in and can't speak clearly or zones out mentally or can't feel emotion, there are no standard tests. The symptom has to be reframed in terms of brain function. The only way to accomplish that reframing is through the elegant choreography of the neurological exam, not with scans, but through the painstaking examination of the patient. Every gesture, every movement, every inflection of speech, every reflex, all of these point to the precise location of the problem in the nervous system, and to its cause. The physician's art is to synthesize the symptoms and signs in a larger framework of the patient and the structure of the nervous system. The tests are merely confirmatory, all the way from carpal tunnel syndrome to brain death.

But for now, the team is still learning about the brain, if not the entire nervous system, observing up close the ways it can malfunction. In time they will figure out how to interact with the person inside the brain, with the brain inside the person. They will find their comfort level with the families—how to give good news, how to give bad news. Someday, they will be the ones who have to tell a patient that nothing will ever be the same.

"You have a brain tumor."

"You have motor neuron disease."

"You have Parkinson's."

"You've just had an irreversible stroke."

What does it mean to be a doctor to these patients? More importantly, what does it mean to be the patient faced with these seismic problems, and how is a connection made with the physician who embodies the knowledge that can make it better?

This book is about the process and progress of my craft. Like clinical neurology itself, it proceeds on a case-by-case basis. The stories are real, the dialogue mostly verbatim, and although the identities of almost all the patients are disguised, the challenges and issues facing the neurologist in a teaching hospital are on full display.

Where does it begin? It begins with a bowling alley manager who gets confused, with a right fielder who starts spouting gibberish, with an undergraduate who suddenly becomes psychotic, with a salesman who drives around and around a traffic rotary, unable to get off, with a college quarterback who can't stop calling the same play, with a psychiatric social worker who notices a tremor in her pinkie finger, with an ex-athlete who cannot grasp the pull tabs on his newborn daughter's diaper, with an Irishman who slips on the ice and cracks his head. They wind up here, on the inpatient ward or in the neuro-intensive care unit, passing through in a parade that never ceases to challenge, astound, and enlighten me.

Where does it end? It doesn't.

Dr. Vandermeer, one of the last survivors of the generation that trained me, sits up on 10 West with a brain tumor the size of a lemon. Gary the programmer is having electrical leads affixed to his head on 10 East. On 10 North, Dennis of the butterfly glioma, along with his girlfriend, have decided to put off hospice for a few more weeks, and they are preparing to go home. Speed rounds have ended, and I am in the elevator, bearing coffee, on my way to 10 South, where Hannah is waiting, patient list in hand. When she sees me, she grabs the hospital ID on the lanyard around her neck, sweeps it over a wall-mounted detector, and the doors to the neuro-ward, like all hospital doors, swing open in counterpoint. As we walk through, she exchanges the patient list for the coffee cup I hold out to her, and I ask the question I have asked countless times:

"So what have we got?"

1

Six Improbable Things
Before Breakfast

Arrivals, departures, and delays on the ward

On the third of July, a day after a routine colonoscopy, Vincent Talma was playing right field in a company softball game. A short, intense man with thick gray hair and a perpetual scowl, he did not look as though he was having fun, or even capable of having any fun. Whenever a teammate said something amusing or cracked a joke, Vincent would laugh without smiling, as if to say, "funny, funny, *ha, ha*." When he disagreed with a call by the umpire, he would throw up his arms in disbelief, kick the dirt, and swear under his breath, not for show or for the approbation of his teammates, but out of real anger and disgust. No one called him Vinnie, few called him Vince, and when he stood at the plate, none of his teammates dared to cheer him on by name.

As the game wound into the late innings, Vincent's behavior began to change, subtly at first, then dramatically. By the time he was dropped off at his house, his wife was startled to see a bemused look on his face, an air of innocence in place of his usual gruffness. He gave one-word answers to her questions, avoided eye contact, and seemed quite unlike himself. He was smiling too much.

"Are you okay?" she asked.

"Sure, fine," he replied.

"Did you win?"

"Fine, yes."

"Did something happen?"

"Fine."

The more she persisted, the more Vincent perseverated.

"Did anything happen at the game?"

"Fine, yeah fine," he mumbled with a sheepish smile.

She called their primary care physician, who told her to get him to the emergency room immediately.

"Vincent, we need to go," she said.

"Fine. Okay." Still smiling.

At East Shore Hospital an MRI showed an ambiguous blotch on the left frontal lobe of Vincent's brain, and at the suggestion of one of his sons, a pediatrician, the family requested a transfer to us. He arrived sometime around 10:00 that morning and was brought up to the ward.

A week earlier, Cindy Song, a sophomore at Boston College, had started acting a bit withdrawn. Her roommate was concerned enough to call Cindy's sister. The first phone call was not too worrisome. "Not a big deal," the sister said. "She gets that way. Just give her time. She'll be okay." The next call could not be taken so lightly.

By morning, Cindy wouldn't leave her room, and would not or could not tell her roommate why. Alternately anxious and distracted, uncharacteristically morose and sullen, she spent the day in bed. That evening she refused to eat, and her roommate made the second call, this one to Cindy's mother, a first-generation Korean immigrant. Despite the language difficulty, there was no mistaking the concern in the roommate's voice. Cindy's mother took the next commuter train from Framingham, exited at Yawkey Station, took the Green Line out to Chestnut Hill, walked up the steep hill from the terminal, past the

Gothic spire of Gasson Hall, and down the long, winding road to the dorms. When she got to Cindy's room and sat down in front of her daughter, all she got back was a blank stare focused on the wall behind her. Her daughter's eyes were wide open and her pupils dilated. She was shivering mildly and sweating all over. Finally, she spoke.

"Mom, they've been after me for weeks, creeping in through the cinderblocks, taking my clothes off."

"What are you talking about, honey?"

"My clothes, my clothes," she said desperately, "can't you see them?"

Like all universities, Boston College has a health center that provides minimal services overnight, on holidays, on weekends, and during the summer, relying on referrals to local emergency rooms for anything serious. The after-hours nurse, who was used to such things, assumed that Cindy had been using recreational drugs and was "just flipping out." Nothing unusual as far as the nurse was concerned, but Cindy's mother was outraged. Convinced simply from cultural experience that there were no drugs involved, she would not let that stand. Cindy was so jittery and sweaty that the nurse gave in and called an ambulance to take her to the Brookline Hospital emergency room. Once there, Cindy remained agitated, stopped responding to questions, and started thrashing, as though reacting to hallucinated visions. This prompted a round of phone calls to the eight local psychiatric hospitals to see if there was a bed for an acutely psychotic young woman. Such beds are hard to come by, and it took a hard sell by the emergency room doctor to secure the promise of one by the next afternoon, "if you could just hang onto her and give her Haldol in the meantime."

As daylight broke, Cindy was close to berserk. Her arms had to be restrained with straps, and she was soaking up tranquilizers like a sponge. Finally, the ambulance came to bring her to the psychiatric hospital. After a short interlude of relative calm, the psych nurses became alarmed when Cindy's jitteriness escalated into full-blown myoclonus—arms and legs flinging up off the bed, her head jerking back violently. Her pupils were huge. If it had been a drug overdose, they realized, this would have abated by now. Instead, the hallucinations continued, and Cindy was excessively restless and sweaty. I got a call at about 9:30 a.m.

"Is she salivating like she has rabies?" I asked the psychiatry resident.

"Yes, like a dog," was the reply.

"You'd better send her over."

By the time Vincent Talma and Cindy Song had settled in at the Brigham, Arwen Cleary had been there for four days. She came by ambulance on the morning of July 1, and was admitted to neurological intensive care from the Emergency Department later that evening. Of the three cases, hers was the least clear-cut, the most troubling, and one that had the potential to become an absolute shambles. According to her medical records, her problems had begun two years earlier, when she showed up at a central Massachusetts hospital with disabling nausea, difficulty walking, and vomiting.

Arwen Cleary had been a professional figure skater as a teenager, had retired from the Ice Capades upon its dissolution in 1995, had then raised three children, gotten divorced, and moved with her two younger children to a ranch house in Leominster, a distant suburb, where she worked part-time at a local health club. Her medical history was unremarkable: once a smoker, she had quit ten years earlier. Her travels had taken her no place more exotic than Bermuda and no more distant than Orlando. Her only hospitalizations to that point had been in maternity wards. She was remarkably fit and in seemingly good cardiovascular health, if judged only by her appearance and vital signs. But shortly after a visit to a chiropractor, she had suffered a vertebral artery dissection, a form of stroke.

Chiropractic neck adjustments are not a common cause of stroke (maybe one in every twenty thousand treatments produces one), but the high rotary force involved, one with just the right vector and amplitude, can strip off the inner layer of a blood vessel, causing it to tear and collapse into the channel, impeding the flow of arterial blood to the brain. At her local hospital, Ms. Cleary was started on a blood thinner, and after a long inpatient and rehab stay, she recovered her motor skills and balance, and was sent home.

All went well for two years, until she returned to the hospital with sudden right facial drooping and difficulty finding words, sure signs of another stroke, but this time a stroke of a very different kind. A portion of one of the language centers of her brain had been deprived of its blood supply. Her speech was now noticeably impaired. Within a few days, she showed signs of improvement, and was again discharged on a blood thinner.

Ten weeks later, to her infinite frustration, it happened yet again, and she arrived at the same hospital in the middle of the night with another language problem, this time even more pronounced, as well as right arm weakness. The scans now showed that several other blood vessels had been stopped up, causing a scattering of new strokes. At that point her doctors became even more worried. Why would this be happening in someone so young? But they could locate neither a cause nor a source. They subjected her to exhaustive tests, the usual suspects for stroke were rounded up, an echocardiogram was ordered, and she was given a portable heart monitor. Everything came back normal. It was decided that the previous chiropractic stroke (the dissection) was unrelated to her current problem. Among the staff, the consensus was: "We're going to need a bigger boat." So they sent her to us.

———

There is an old joke among stand-up comics that goes: "Dying is easy, comedy is hard." If we were as inner-directed as comedians, we neurologists might say, "Trauma is easy, neurology is hard." Every one of our patients has, in effect, fallen into a hole, and it's our job is to get them out again.

In *Alice's Adventures in Wonderland*, Alice jumps into a rabbit hole and finds herself in a bizarre realm in which nothing is what it seems, where everything bears little relation to the outside world. It is a place where, as the Red Queen mentions to Alice, it helps to believe six impossible things before breakfast. Unlike the Queen, I have no need to believe six impossible things before breakfast because I know that, on any given day, I will be confronted by at least six improbable things before lunch: a smiling man whose speech difficulties seemed to have been brought on by a colonoscopy, a thrashing young woman whose psychosis seemed to come out of nowhere, a figure skater with a slow-fuse time bomb in her body that was knocking off her faculties one by one. The first of these, I should note, was indeed impossible, and I didn't believe it for a second, but the next two were quite possible, and by the end of the morning, I would encounter at least three more improbabilities: a woman who could only be cured by a hole in the head, a case of amnesia brought on by sex, and a man who was adamant that I was two very different doctors.

We treat people with seemingly implausible ailments all of the time. Each day they show up in a predictable parade of signs, symptoms, and diseases: an embolus, a glioma, a hydrocephalus; a bleed, a seizure, a hemiplegia. That's how the residents refer to the cases, as in: "Let's go see the basilar thrombosis on 10 East." When viewed in terms of actual patients, however, no day is quite like any other. After the bedside visit, the thrombosis suddenly has a name, the glioma has a wife and children, the hydrocephalus writes a column for a well-known business journal. Our coed suffering from psychosis turned out to be a Rhodes Scholarship candidate, the case of multiple strokes became a charming woman who had competed in the Junior Olympics, and the man for whom a smile was a troubling symptom owned a personal empire of six Verizon wireless stores.

"Good morning, Mr. Talma," Hannah said, "do you remember me?"

"Yes, good, good, fine," Vincent replied. He was sitting up in bed, watching television with a smile of bemused innocence. Vincent Talma was a picture of contentment. His room on the tenth floor of the hospital tower commanded an outstanding view of Fort Hill Park in Boston's Roxbury section, but Vincent took no notice. Along with twenty-nine of our other patients, he had been waiting for a visit from the neurology team on their morning speed rounds.

Hannah was in charge. Her service, the culmination of three years as a neurological resident, had started a week before I came on board. A "service" involves running the neurology inpatient ward, admitting and discharging the patients, and directing a team consisting of three junior residents, two medical students, and a physician's assistant—a cohort that could barely squeeze into Vincent's curtained-off half of the room.

My colleagues and I had some doubts about Hannah when she first came to the program three years earlier. The most superficial of these doubts focused on her style of dress. In a profession where sartorial flair is an unexpected and somewhat suspect concept, Hannah's clogs, leggings, and wraps seemed needlessly exotic, and sowed uneasiness among the Dockers, Skechers, and scrubs crowd. Perhaps even more alienating was the fact that Hannah did not drive a car, and instead rode her bike from her apartment in Boston's North End to the Brigham, usually well before the sun rose or long after it had set, in any kind of weather short of a blizzard. Such stoicism flew in the face of the unhealthy lifestyle adopted by most of the residents and teaching faculty, who tend to favor pastries over granola, Coke over water, and elevators over stairs.

I could see that over the course of the previous week, Hannah had begun the transition from resident to full-fledged physician. I could see it in her bearing, in the assertive physicality with which she carried out her examinations, in the firmness of her tone with some of the more difficult patients, and in the controlled sympathy she adopted in family meetings when she had to deliver bad news. She had turned out to be one of our strongest clinicians.

Although she hails from the Midwest, Hannah Ross has a northern

European flair, somewhat Dutch, in that she is tall, lithe, wears fashionably businesslike glasses, and seems indifferent to the possibility that anyone might appreciate the effort she has made in choosing her look, probably because the effort is now merely a habit. She moves swiftly from room to room, from pod to pod, from the nurses' station to the rolling laptop cart, where she displays an instantaneous command of electronic medical records, and can bring up an MRI scan and zoom in on a tumor or a cerebral hemorrhage with no wasted effort.

"What are you watching?" Hannah asked Vincent, in an inflection she would later inform me was Kansan rather than Missourian.

"The Bunkers."

"Do you mean *All in the Family*?"

"Yes, yes, . . . the Bunk . . . Yes."

Vincent's form of speech difficulty, known as Wernicke's aphasia, sounds like gibberish, but not pure nonsense. It can include halting phrases that almost make sense, echolalia (repeating someone else's just-used words), perseveration (giving the same answer to a succession of different questions), and play association (cracking wise). While he knew the answers to many of our questions, most of his responses didn't come out quite right, yet he seemed unaware and unconcerned.

"What's your name?" Hannah said.

"Vincent."

"Good. Where are we? What place is this?"

"Vincent . . . uh, yeah . . . Vince."

"What day is it?"

"Avince . . . Vince."

"Okay. Look at my hand. Now follow my thumb."

"Gee, you're so dumb."

Gilbert, the medical student who had made the initial exam, recorded this as "orientation times one."

"To one what?" I later asked him.

"To himself," he said.

"Have you ever met a patient who wasn't?"

"I don't think so."

"No, you haven't. It doesn't exist."

The phrase *A and O times three* means "awake, oriented to self, oriented to place, and oriented to time." Some people add a fourth: oriented to situation. The problem is that everybody is "oriented times one" unless they are hysterical or dead.

Vincent knew who he was. He was sharp enough to find himself amusing. Did his colonoscopy earlier in the week bring this on, or, more to the point, did the anesthesia bring it on? My guess is that it was just a coincidence. A straw poll of the team leaned toward a diagnosis of tumor, possibly stroke, maybe a seizure, but they were basing their guesses on Vincent's MRI. I had seen the scans and knew they did not hold the answer. On the other hand, Vincent's wife, who was sitting in an armchair at the foot of his bed, did.

"He had a bad headache from the beginning," she told me, "and a fever." The residents had neglected to mention this, but it was important.

"How about a virus?" I suggested. "I think this is probably an infection." Herpes encephalitis was my hunch. It would connect the headache and low-grade fever, neither of which fit with a tumor or a stroke. "Ignore the scan for now," I told Hannah. "When there's nothing obvious there, it can be a distraction. Stick with the patient's story and the bedside exam."

We started him on acyclovir, an antiviral medication, and he soon improved. Five days later, Vince was discharged, talking normally again, and, for better or worse, just like his old self.

"I just ran into your Mr. Talma in the elevator lobby." Elliott, a colleague who seems to keep closer tabs on my patients than I do, had buttonholed me in the corridor outside of the ward. "When I gave him a shout-out," he said, "you'd think I'd asked him to put up bail for the Unabomber. The guy comes in here a pussycat, and when you finish with him he's Mr. What's-It-To-You-Pal. No more smiles, no more jokes. What did you do to him?"

"We cured him," I said. "Apparently, that's his baseline. I told his wife that if he started being nice to her again she should bring him back in immediately."

I was out on the ward at about 9:30 that morning when the call about Cindy Song arrived from the other hospital.

"Is she salivating like she has rabies?" That was my first question, and would turn out to be my only one.

"Yes, like a dog," was the reply.

"Holy cow!" I said. "It's an ovarian teratoma. You'd better send her over." It was a snap diagnosis, possibly wrong, but there was no harm in raising on a pair of aces. I had a pretty good idea what the other cards would be: memory deficits, gooseflesh, a high heart rate, and no family history of psychosis. The drooling alone was a tip-off.

A teratoma is an unusual tumor that contains cells from the brain, teeth, hair, skin, and bone. Most teratomas are harmless, but they have the potential to wreak havoc by causing encephalitis. When you see it, the syndrome is unmistakable: an ovarian teratoma stimulates an antibody that will produce the very ensemble of symptoms that were described to me over the phone.

Two hours later, when she was wheeled into the ICU, Cindy looked toxically ill, with a heart rate of 135 beats per minute and blood pressure of 160/90. She was sweating, salivating, and shivering wildly. Her eyes were wide open but she was by now entirely unresponsive. Her jittery limbs seemed as if they wanted to convulse. Joelle, the senior ICU resident, Hannah's counterpart down on the ninth floor, immediately intubated her.

The toxicology screen from the other hospital was negative, so I called the gynecology service to get an emergency ultrasound of Cindy's pelvis. They thought I was crazy. Moreover, I insisted that they do it transvaginally in order to get a good look at the ovaries. An ovarian teratoma can produce memory loss, seizures, and confusion—what neurologists call "limbic encephalitis," or sometimes the "Ophelia syndrome" (not for Hamlet's beloved, but for the daughter of the neurologist who described a similar condition). The psychotic symptoms are due to autoimmune antibodies that attach to a receptor in the brain, where they simulate the effects of PCP (aka "angel dust" or "wet"). When that receptor became blocked in Cindy's brain—when the antibodies hit their target—all of her symptoms became manifest. She went nuts.

"Remove her ovary?" the gynecologist said.

"Right. Do you see that cyst on the ultrasound? It's not so benign."

I had to insist that there was now no doubt about it: the ovary-brain connection. First—"Who would have thought?" Then—"What do you know? It's a real thing." Eventually, both the resident and the attending gynecologist were convinced, and they were comfortable knowing that Cindy could still have children with her remaining ovary.

This was a rare, rare thing. No one fully understands it, but I know it clinically when I see it, or even hear it over the phone, because I collect arcana. If the problem is properly framed, there are very few other things it could be. It took a bit of cajoling, but in the end, they removed her ovary. The sweating, the salivating, and the wild swings in blood pressure were gone within hours. Her psychosis resolved within days.

Back on ten, Arwen Cleary, our ice skater with the multiple strokes, had gone deeper into the rabbit hole than anyone else on the ward, and I wasn't confident that we could pull her out of it. According to the notes in her chart, she had by now had three separate strokes, clearly visible on MRI scans, in addition to the vertebral dissection from her neck manipulation. An angiogram had been interpreted as showing vasculitis, an inflammation of the blood vessels. She had a subplural lesion in her left lung, according to the pulmonary specialist. She had a low platelet count, according to the hematologist. "The patient uses humor to cope with her situation," according to the social worker, and on and on for thirty pages of cut-and-pasted notes from more than two dozen doctors who had examined her over the past two months: too many specialists weighing in with too many disconnected analyses, not adding up to a complete picture. Most of her file consisted of blind alleys and misinterpretations.

She was in rough shape, virtually blind in her right field of vision, and now aphasic. What worried me was that she didn't have any reserve left, and any little chip-shot stroke was going to be a disaster. The next one, I was convinced, could wipe her out.

"I don't see any vasculitis here," I told Hannah. The low platelet count, which if anything would tend to protect against clotting and stroke, was another red herring. "I think the thing to do is just start from scratch. Something is missing. We've got a new team, so just make like she's being seen for the first time, make believe she hasn't been worked up, fill in all the holes. I think there's a single origin of these multiple emboli. That's what it sounds like, that's what it looks like. There's something upstream that's flicking off debris into the blood vessels of the brain, and we just haven't found it yet. If you told me she had a myxoma, I wouldn't be surprised."

Something had to be giving off small flecks that lodged on the walls of blood vessels, effectively narrowing them. That was what had caused the strokes, and that was what had been misinterpreted as vasculitis. It was happening now, and would continue to happen, and the most logical source for the flecks had to be a thrombus (a clot of some kind), a tumor (a myxoma or fibroelastoma), or a bacterial growth due to an infection, probably in or near one of the valves of her heart. Yet my residents insisted that there was nothing wrong on the echocardiogram. After sifting through the case file, we finally got around to visiting her.

"Hello. This is the neurology team. How are you?"

"Not so hot."

Unlike most of our patients, Arwen Cleary did not look sick. Not only did she look physically fit, but physically vibrant. Rather than sagging into the hospital bed, she balanced on it like a coiled spring, ready to jump out of it if necessary. At the same time she was shy, somewhat abashed at being here. She had had no visitors for over a day, possibly because she did not want her children to see her like this, or, more accurately, to *hear* her like this, for although she could talk, she could only do so with halting fluency, mostly in monosyllables. She struggled and usually failed to come up with the longer words that best expressed her thoughts.

"I know, it's tough, not being able to express yourself easily."

"Oh, yeah. I'm off . . . Oh, my gosh!"

"The dissections . . . I read in the chart that that happened after you got chiropractic treatment. Is that true?

"Right. Yes."

"How much time elapsed, between the two."

"It was . . . just a few days."

"You know your spirits have been marvelous despite all this. How are you doing it?"

"I . . . I . . ."

"You stay optimistic."

"You have to." She spoke with an unnatural monotone, somewhat like a deaf person, without accenting any of her words. That was the aphasia. She struggled with all but the simplest responses, and settled for tropes.

"Do you find yourself getting down sometimes?" I asked.

"Some . . . times."

"Are you depressed?"

"No, not depressed . . . just sick of all this."

"Discouraged?"

"Yeah."

"Well, thanks for letting us spend some time with you. We're racking our brains to figure out what's going on."

Stroke offers the most precise and restricted indicator of damage to the brain that nature produces, and therefore allows an understanding of brain function like no other disease. It is highly "readable," and reading strokes reveals a tremendous amount about the nervous system. One of my professors used to say that the residents learn neurology stroke by stroke. But it is not a simple thing.

Six thousand people have a stroke in the United States every day. The numbers are overwhelming. The country has a stroke belt which runs from North Carolina right through Oklahoma. There are genetic factors and dietary ones. Scandinavians have the fewest strokes, the Japanese have the most. There are at least three broad categories of stroke: one involving blocked blood vessels, another involving bleeding into the brain, and a third—an aneurysm—involving a ruptured bulge in a blood vessel. Although these are all called strokes, they are as different from each other as hepatitis is from gall bladder disease (both of which give you jaundice). And their treatments are entirely different.

My experience told me that Arwen Cleary's echocardiogram had missed something, not just once, but twice. I turned to Hannah after we had left the room, and asked her this crucial question: "Can you put your finger on what's different in this case?" She replied that it was the angiogram, which showed the alarming number of constricted blood vessels and cutoffs in the cerebral arteries.

"No," I said. "It's the recurrence of many, many small strokes *over time*. That's what's different. You have to think about what could cause this. There must be a cardiac source for the emboli. Do the echo over. It was wrong. If it doesn't show something abnormal on one of her heart valves or in one of the chambers, I'll eat my hat."

Medical textbooks teach you what tests to do to make a diagnosis, but they do not dwell on the simple reality that humans are interpreting the tests. Hannah ordered the echocardiogram yet again. This would be the third one. The hospital would have to eat the cost. The cardiology fellows initially balked, but when we showed them the echocardiogram from the other hospital, they came around. It was incomplete. By way of proof, the repeat TEE showed a mass sitting on her mitral valve—a papillary fibroelastoma, the second most common benign tumor of the mitral valves, and one which took the shape, as I had predicted, of a peduncular (or branched) growth. I wasn't entirely right. The cardiologists did not think it was consistent with an atrial myxoma, a very different kind of tumor, but at least I didn't have to eat my hat.

The course of treatment seemed clear to me: the tumor would have to be removed as soon as possible, before another stroke occurred, and her mitral valve would have to be replaced. Although this seemed to be an answer, when I ran it by the head of cardiothoracic surgery, he balked. "There's too high a risk she'll have a cerebral hemorrhage on the heart pump. We'll have to wait six weeks so her last stroke won't turn into a brain hemorrhage."

———

Alice in Wonderland is an absurdistan story. Beyond fantasy, it's ridiculous. That's neurology in a nutshell. Your patient disappears down a rabbit hole. You've got to do something. You can't just sit there, so you go down the hole after the patient. Sometimes you can do it right away: you go to the gynecologist and say, "Take out her ovary," and that gets her out of the hole. It may not get her back out the same hole she went in, but in a case like Cindy Song's, it gets her out relatively quickly. Same with Vincent Talma. We brought him out, not quite as good as new, with a slight speech deficit that most people wouldn't even notice, but we got him out. With Arwen Cleary it would be a longer journey. Although it went unmentioned at morning rounds, her case would offer a sobering reminder that there are significant limits to our knowledge of diseases of the human nervous system. Anyone expecting a clean resolution and a quick turnaround was in for a disappointment.

Arwen Cleary remained on the ward for five weeks. She did suffer a cerebral hemorrhage, but it resolved with almost no consequence. The fibroelastoma that looked so ominous on her TEE somehow disappeared, or perhaps it wasn't a fibroelastoma to begin with. Because of the blood thinner we gave her, she suffered no more strokes. Eventually she went to rehab, and from there she went home. She did not get the heart surgery. She would have to stay on the blood thinners for the rest of her life, and I may never be able to say what caused her problem, or whether it was still a problem, until she suffers the big stroke that wrecks her.

Six months later, she came to the outpatient clinic, her speech much better, but still frustratingly limited. Her vision had not fully returned. She was making very slow improvement.

"What's your account of what's going on?" I asked her.

"I think it stinks."

"Are you optimistic you're going to get better?"

"Yup."

"Can you tell me what you think of your experiences in the hospital?"

"I feel like I've never left here. Some days, I can wake up and say, 'Oh, it's going to be a very good day.' Then, it sucks."

"What kind of person were you before all these strokes, and what kind of person are you now?"

"I . . . I was always on the go. Four hours of sleep. Upbeat. Dancing . . . Yeah. I used to walk. Now I can't see clear."

"That's because you're missing the right side of the world. You might not be aware of it, but your vision on the right is diminished because of one of the very early strokes. Are you a different person now?"

"Once I'm home, I'm good. Like, I just . . ." She trailed off.

"Are you very weepy?"

"Kind of."

"I think this will settle down and there will be a new equilibrium where you're better than you are now. And I hear you, that the dizziness is what's driving you crazy. I know it's frustrating, but your kind of case can't be solved by a book, or it would have been solved by now."

"It's not simple," I tell Gilbert, the third-year medical student, "almost nothing is routine, but if at the right moment you can combine experience, logic, and leaps of imagination, you'll get your patients where they need to go."

That's the pitch. Gilbert has to decide on which specialty to choose by the end of the year, and that is the extent of the effort I will make to sell him on clinical neurology. Rounding on the ward will either appeal to him or it won't. It's not for everyone. Among the residents on the team, who have already chosen neurology, some will concentrate on research and try to find the causes and cures of Parkinson's disease, Alzheimer's, or multiple sclerosis. Some will go into pediatric neurology. Others will become epilepsy or stroke specialists, some will go into psychiatry. But a few special ones, like Hannah, will carry on the clinical tradition, one case at a time.

Back on the ward, she comes up to me with the patient list. I am waiting at the nurses' station with the rest of the team. "Elliott thought you ought to see this lady first before we make rounds," she tells me. "Her name is Mrs. G, and she's making me nervous."

"Why?"

"She's the lady with the hydrocephalus." In other words, she has too much water in the cavities of her brain, a serious problem.

"Lead on," I reply. "I'm at your service."

2

Like a Hole in the Head

———————

Where baseball and neurology converge
in a game-saving, over-the-shoulder catch

On Tuesday, at Chief's Rounds, a weekly ritual that takes place in the department library, Elliott handed me two tickets to the Red Sox–Mariners game, a 7:30 p.m. start. Like every home game, this one was sold out, but Elliott has season tickets. Although temperatures had already hit the 90s, by game time they would be in the low 80s, then back down into the 70s by the stretch. With clear skies and a crescent moon, it promised to be the kind of idyllic Boston evening that you dream about in mid-February, and feel entitled to by early July.

Elliott is an odd duck. No one else in the department wears monogrammed shirts with cuff links, sports a Patek Philippe watch, or knows as much about pari-mutuel betting. He started in private practice and was such a good neurologist that we hired him into the academic group, where he had a meteoric rise from instructor to professor. Not one to gossip, he nonetheless seems to have a wealth of inside information. He knows what the residents are up to, what the administration is thinking, who is the next to be fired. He is a classically handsome man in the *GQ* sense, square-jawed, still maintaining a wrestler's build into his midforties. He went to a state university and to a less-than-name-brand medical school, and is unimpressed to the point of indifference by his current affiliation with Harvard University. He comes and goes as if he has something better to do, and apparently he did have something better to do than Chief's Rounds, because when I turned to thank him, he was gone. The tickets were three rows behind the visiting team's dugout. I would be wrapped up here by 5:00 p.m., I was thinking, unless something went terribly wrong. Which it did.

Late in the afternoon, I was pushing the rolling cart with the patients' charts around the semicircular nurses' station on the tenth-floor ward. As I put the finishing touches on the last note in the last chart, a floor nurse rushed out of room 41 West and asked me if the team was going back in to see Mrs. G, the woman admitted last night from the intensive care unit, the very patient that Hannah and Elliott had been so concerned about. Mrs. G was Sofia Gyftopoulos, and her condition—hydrocephalus, also known as water on the brain, accompanied by a history of meningitis—was serious, but not critical.

"No, we saw her an hour ago on our afternoon walk-rounds," I said, "and she seemed fine except for a headache. Would you like me to take another look?"

"Yes, I think her breathing is shallow."

If not for the unusually high number of admissions over the holiday weekend, I would have been home, resting up for the game. It would have been Hannah's problem instead of mine. But I go where I'm needed.

I walked in and briefly greeted Mrs. G's husband Nick, who was sitting at the foot of the bed reading *Entertainment Weekly*, not looking at all concerned. Neither was Mrs. G, but then she wasn't conscious, and her breathing was forced. I called her name and shook her shoulder, but was unable to rouse her. I checked her pulse, found it to be thready and barely detectable, and when I pried open her right eyelid, I saw that the pupil was enlarged and had lost its natural reflex of constricting in response to light. We call this a blown pupil, and it is a neurological sign that the brain is about to collapse. Nick was on his feet, *Entertainment Weekly* was on the floor, and the nurse was standing about three feet behind me when I turned to her and said in a loud but controlled voice: "Call a code!"

Within the next two minutes, a dozen people would rush into the room, none of them having a precise idea of what was happening inside Mrs. G's head, why her brain structures were shutting down one by one. While the code team focused on other things, like keeping her heart and lungs going, my job was to get to the root of the problem, and fix it. As for the Red Sox, they would have to start the game without me.

In the clarity of hindsight, Mrs. G's event was entirely predictable: not so much the *when* as the *why*, and the *how*. The stage had been set earlier in the day at morning report, when Hannah informed me that there were thirteen admissions to the neurology ward during the night and only three discharges. Two of our patients were sent to the ICU; one had coded three times overnight, and the patient list, usually no more than a page and a half, had lengthened to three full pages. Flavio, a second-year resident and a key member of the team, was stuck in Madrid with a visa problem. To add insult to an entire roster of injuries, an aging rock star, a Boomer legacy act, was scheduled to arrive sometime that morning and be whisked to a pavilion suite on the sixteenth floor, shielded from the prying eyes of the press under a pseudonym. He was going to want our full attention. Which is to say we were short-staffed, overbooked, and had a celebrity admission to boot. It was a potentially dangerous situation, not for our rock star, but for some of our higher-risk patients like Mrs. G. And I had only myself to blame.

For years now, I have scheduled myself for "the service" during the first two weeks in July and the last two weeks of December, meaning that I sign up to serve as the attending physician on the neurology inpatient ward and the neuro-ICU, partly as a favor to those who would rather not work through the big holidays—the Fourth of July and Christmas—and partly to show them up. I don't mind. I'm happy to do it even though these tend to be slow times on the wards. To compensate, I encourage the residents to admit as many patients as they can. "Keep an eye out for interesting cases!" I tell them. "Go walk up and down the sidewalk in front of Au Bon Pain and see if you can spot anyone who can't walk straight."

It was July 3, the third day of my service. I had taken over from Elliott, who was now on neurological consult to the Emergency Department. I had been relieved to learn that I would be getting Hannah as my senior resident. She had started her rotation a week earlier, so she was up to speed on most of the patients. As for the rest of the team, I never know who is scheduled until they show up. Given the situation with Flavio, we had been forced to go to the bullpen and come up with two first-years from Children's Hospital, who would have to be tutored in the Brigham's arcane medical ordering system, written in DOS sometime during the Ford administration.

It should not come as a shock that the daily routine at a teaching hospital does not much resemble the tightly choreographed one-hour dramas that dominate prime-time TV schedules. The sheer numbers of patients, their tendency to pass in and out of our service due to medical rather than plot-driven priorities, restricts our time with each one of them, and sometimes scatters our focus. During those two weeks, Hannah would often tell me that if she simply had the time, she could crack some of our most baffling cases. If not for the paperwork and the bureaucratic overhead, perhaps she could have. But time is a luxury, and sometimes it feels as though our primary function is just to check off boxes on a never-ending punch list. As I said to one of the new residents, "You may get the idea that we're constantly draining the swamp."

That morning, like every morning, the neurology team had gathered in the conference room on the tenth floor of the hospital tower—a cramped, windowless, cluttered way station where the various medical teams convene to discuss their cases, order lab tests and consultations, and steal an occasional power nap. The room is a study in off-whites: an uninviting, fluorescent-lit, purely functional space. Melamine counters and computer workstations run along the right side, white boards dominate the left, and a conference table with office chairs is crammed into the middle. A fifty-inch LCD flat screen mounted on the far wall is used to display scans and test results. There was barely space for eight of us and the food we had brought: enough bagels, doughnuts, muffins, Danish, and coffee to ensure an elevated glycemic index for the next eight hours.

As she handed me the patient list, Hannah informed me that one patient, a Mr. Williams, the man who had coded three times in one night, should never have been sent to the service in the first place, and had kept Elena's hands full. As the overnight resident, Elena had had almost no downtime on her shift, was clearly drained, and was now ninety minutes away from the end of her mandated maximum of fourteen consecutive hours. For the next hour and a half she would complete the "handoff," an unfortunate, policy-driven ritual in which the doctor who knows the most about the new patients, the only one who has actually met and examined them, presents the essential details of the cases to the rest of the team, and then leaves, often unwillingly. If all goes well, the baton passes smoothly and we don't break stride. But every now and then we miss a step, or drop the damn thing. It would take us over two hours to run the patient list that morning, and during that time we would squeeze in speed rounds early so that Elena could go home and get some rest. "Just the news," I kept telling the residents, "not the weather."

Among the new admissions, one patient worried me: Mrs. Gyftopoulos, a fifty-year-old mother of three. Was she stable enough to be here? I didn't think so. As we headed off on rounds, I said to Hannah, "We've got too many patients. This is getting a little unsafe."

Elsewhere on the ward, we had Mrs. Newlin, a ninety-year-old woman with such explosive headaches that Elliott suggested putting her on a terrorist watch list. We had two drug abusers sharing a room, one a burnt-out addict with horrible teeth (he had thrown away his toothbrush because it reminded him of his wife, who had left him six months ago), the other a pure drug seeker in his early twenties, who I had admitted from the outpatient clinic after he exhibited such excruciating sciatic pain that I felt I had no choice. Later in the day, Elliott informed me that he had seen the guy down in the lobby eating a fruit cup while perched on the edge of an armchair. "No evident signs of distress, in fact quite the opposite," Elliott informed me. It would take a concerted team effort and a final bribe of intravenous Dilaudid just to get rid of the kid.

We had our aging rocker in the luxury suite, accompanied by a small retinue, including a personal trainer, an aromatherapist, and a bodyguard (strictly against hospital rules). In turn, I brought my own entourage of residents and med students, most of whom had never heard of the guy, having been born well after his last album had gone gold. Back on the ward, we had Doc Vandermeer, with the lemon-sized tumor in his frontal lobe. As patrician a man as one might meet, he had the endearing habit (or infuriating one, depending on your outlook) of never saying anything directly that could be couched in clever circumlocution. When Hannah asked him if he needed anything to make him comfortable, he replied, "No, but my equanimity and support services appear to be compromised in a telling way, and to get back my life would be gratifying." It was already clear to me that he would not be getting much of his life back unless we removed his tumor, and even then it was not a sure thing.

Nearby we had a larger-than-life, left-leaning economist and activist who, with very good reasons that I worked hard to deflect, harbored a destabilizing conviction that she was experiencing the onset of Lou Gehrig's disease. Next door, a Boston firefighter was coming to grips with the fact that he had a glioblastoma, a rapidly growing, incurable, and inoperable tumor that would kill him within six months.

Down the hall, a sham epileptic was manifesting theatrical seizures that registered nary a blip on the EEG. And scattered here and there were a variety of ischemic and hemorrhagic strokes, both catastrophic and benign, one a Wernicke's aphasia and one a Broca's aphasia, the first of which plays havoc with word selection, the second with speech production.

Then there was Mrs. G herself. The details of her case were presented at morning conference by Callie, a second-year resident with an MD and PhD from Yale, who would have looked more at home on Melrose Place than on Chapel Street. Callie spoke with an up-lilting intonation, and delivered the patient history with a curious mixture of arcane medical nomenclature and L.A. street slang.

"She's a forty-nine-year-old woman with a history of basilar predominant severe lepto-meningial inflammatory syndrome that's been worked up up the wazoo. At baseline, she can kind of shuffle. She has some intermittent diplopia and dysarthria, and she's presenting with increased difficulty with walking, and also a gnarly occipital headache. She can't walk without two people assisting her. For the last six days she's had to concentrate a lot more than usual. Of note, in addition to this lepto-meningial inflammatory syndrome, she's had chronic hydrocephalus."

More plainly put, Mrs. G had serious problems, had been in and out of the hospital, and was nearing a kind of crisis. Her scans, when Hannah put them up on the flat screen, looked so unusual as to be alarming.

"What's going on there?" I asked. Clearly, we were coming into the middle of a very complicated case. I was immediately bothered by the pear shape of the ventricles, the fluid-filled cavities in the middle of the brain. I had seen this before in brains in which the spinal fluid was under very high pressure. The team seemed to appreciate that the ventricles were enlarged, but did not pick up on the fact that they were also under substantial tension. I took this as a cue to look for worrisome signs when we finally got around to the bedside.

"She shouldn't be here on the ward, the ICU would be better," I said. "She has massive hydrocephalus. She's not going to able to walk soon."

"So we essentially cut right to the chase?"

"Right. There's no time to make this diagnosis. She needs a shunt ASAP." A shunt is an internal one-way drainage valve that relieves and controls the internal pressure of cerebrospinal fluid.

"Neurosurg needed a place to put her while they figured this out," Callie said.

"Well I can figure it out for them: send her back to the ICU, have them do a temporary, external shunt, get some cerebrospinal fluid to test, get her walking, make a diagnosis, and *then* do a permanent shunt. We need to prevent her from getting demented and being wheelchairbound. I know this is meta-level stuff, but we should keep our eye on the ball here."

I once had a professor who used to say you should never joke with your patients, and you should never return their jocularities. He said it was a bad, bad mistake. I have never followed that advice, although it might be a good rule to apply with the residents. In hindsight, I should have stayed on message. Instead, I asked if Mrs. G had any unusual exposures. "She has a Greek name. Does she work any place special? The last case of this I saw was a guard at the Egyptian room at the Museum of Fine Arts. He got Nocardia from the mummy." Nocardia is a bacterium, typically found in soil that, if inhaled, can cause a slowly progressing pneumonia. In rare cases, it can cause an inflammation of the meninges, the brain's protective sac, and lead to hydrocephalus. The museum is just down the street from the hospital. It was an interesting case, but didn't shed any light on the matter at hand.

"Oh, that's awesome," Callie said. "Actually, I don't know what her exposures are, her mummy exposures."

As entertaining as this was, my little anecdote diverted the team's attention from the very point I wanted to drive home: that this woman should be sent back to the ICU or even straight to the operating room.

We needed to keep a close eye on her to make sure that the pressure in her brain did not cause her to crash. But in a setting in which thirty cases are discussed within two hours, no patient gets more than fifteen minutes of fame, making it doubly imperative that we stick to relevant facts. Instead, I took my eye off the ball, and the rest of the team did, too.

When we finally got to see Mrs. G on rounds that morning, I was impressed by her odd combination of mental dullness accompanied by considerable head pain. The bluntness told me that her hydrocephalus was pressing on the frontal lobes from the inside, and winding down the engine that makes the nervous system spin. The pain that caused her to intermittently clutch her head and neck suggested that the pressure inside her skull was not only very high, but rising.

The skull is like a fixed container. If pressures within the skull are unevenly distributed, there is a breaking point at which the entire brain gets squeezed downward like a plunger, compressing the brain stem. This is precisely what happened when Mrs. G blew her pupil later that afternoon, resulting in the code blue. I just happened to be outside the door to her room when the problem came to a head (so to speak).

Once a code blue is called, it takes a minute or two for the team to materialize from the other pods and from the ICU a floor below. In that brief interlude the nursing staff swings into action. Three nurses rushed in to help move the patient from her recliner onto the bed. As we prepared to roll her onto a board for chest compressions, I wiped the sweat off her back so the defibrillation pads would stick, then pushed the triangular mask of the Ambu bag around her mouth and nose.

"She's going to need mannitol," I said, and as I came out of the door I ran into Hannah, who seemed genuinely surprised when I told her that Mrs. G had blown a pupil.

"This woman needs mannitol!" I yelled again over to the desk nurse, and Hannah went to help her get it. Mannitol, a drug administered by IV, is a sugar alcohol used to pull fluid out of the brain in order to reduce the internal pressure. In this case it was just a stop-gap. Mrs. G was going to need more than mannitol.

A code is a highly choreographed performance executed in a small space measuring approximately eight by twelve feet. Among the dozen people who rushed into the room in the next few minutes, each one had a specific part to play, much like the musicians in an ensemble. The code leader, a senior medical resident, is nominally in charge, but because Mrs. G was my patient, I took up a position at the head of the bed and "directed" the code while the senior resident "ran" it. In a sense, I appointed myself guest conductor. I had to be there to provide my perspective, because for a code team, whatever's going on with the brain is secondary. Their primary focus is to restore respiration and circulation—keep her heart and lungs going.

A passerby could be forgiven for mistaking a code blue for an assault. A junior resident was kneeling over the bed, stiff-arming the patient's breast bone to force blood through the heart chambers. On the third pump, I heard the audible crunch of Mrs. G's ribs cracking, a sign that the resident was doing the chest compressions forcefully enough. That was her job. My job was to tell the code leader that the brain was not secondary in this case, that its internal pressure was pushing the brain stem down onto itself, causing the nerve cells that control breathing and heart rate to shut down, and that the only way to resuscitate her would be to reverse that vertical displacement.

At two minutes into the code, the mannitol was being infused, but I lost her pulse at both the carotid and radial arteries, and asked one of the nurses to call down to the neuro-ICU and get a neurosurgery resident with a ventricular tray right away. Hannah, who had placed the intravenous line that delivered the mannitol, would later admit to me what was going through her mind at the time: nothing. She was completely nonplussed. For weeks afterward she agonized over whether, had I not been there, she could have composed herself enough to figure out why the patient had crashed and what needed to be done about it. What Mrs. G needed, literally, was a hole in the head. She needed to have a tube threaded into her brain through which the excess fluid could be drained from her over-pressurized ventricles. The procedure, called an EVD, or external ventricular drain, is done with a hand drill—the eggbeater—which cuts a hole in the skull, through which a straw-like tube is inserted. Both of these items were brought up on a tray by the neurosurgeon, a young Chinese American woman who stood just over five feet tall.

Meanwhile, the code leader took charge and told everyone to quiet down. As one nurse injected epinephrine intravenously, another was kneeling on the bed and pumping Mrs. G's chest. Floor nurses swarmed like ants around poor old Mrs. Newlin, the ninety-year-old with explosive headaches who shared the room, and whisked her away to another pod. The room was filling up with people, and the floor was now littered with about six inches of trash. There were syringe covers, a dozen plastic bags, little cardboard containers underfoot, and four-by-four gauze pads everywhere. The team had administered two shocks through the defibrillator pads, and Mrs. G had briefly regained a pulse and blood pressure, but then lost them again. A third shock was administered.

While the neurosurgeon got set up, I stepped out of the room to talk to Nick, Mrs. G's husband, ushering him to the far wall of the nurses' station, though still in sight of the commotion. He was clearly stunned. I wanted to let him know that I thought we'd be able to bring her back. I asked him if he was doing alright, but his answer was perfunctory. A social worker came by to stay with him.

Back in the room, a surgical assistant shaved and scrubbed a large patch on the right side of Mrs. G's scalp. Hair fell to the floor in clumps, and the area glowed orange with iodine disinfectant. The neurosurgical resident quickly donned a mask, gown, and gloves, and with hands aloft she deftly pirouetted so that I could wind and tie the paper belt on her gown. She poured another half bottle of iodine on the patient's scalp, took out a tape measure, and began to mark Mrs. G's head with a sterile pen. From the bridge of the nose she drew a line up the scalp, measured five centimeters to the side, and made a quick large X. The scalpel came out of its scabbard, and a one-inch linear incision appeared from nowhere down to the skull. She then inserted a small retractor, took the eggbeater drill out of the kit, and pressed down on the back of it. As she turned the handle, bone chips started coming up the drill bit. Ten turns in, and the drill stuck.

Later, Hannah went around saying that I ran the code, but I hadn't. I made some midcourse corrections for the medical resident, and I was cheerleading the neurosurgeon. When the drill got stuck I was about to put on a gown and gloves to give her a hand, but she persisted. Seeing a diminutive Chinese American woman lean in like a rugby player in a scrum, and push with her whole body weight against the skull of a fragile woman was quite a sight. It was getting hot, and I was sweating. The neurosurgical resident was sweating more. She braced herself for one last push and the drill broke through the cranium. I told her that she didn't need to make a perfect trajectory with the catheter since the patient's ventricles were huge, and she could probably hit them from almost any angle. "Don't tunnel it under the skin," I told her, "we don't have time, just stick it right in." But she was in her zone, and she dutifully dragged the ventricular tubing under the scalp as she had been taught, as if we had all the time in the world. Suddenly, a fountain of spinal fluid spurted in a jet past my left ear. I could hear it whizzing by. Everyone at the head of the bed looked incredulously at the wet splash running down the wall behind me. In less than five seconds, Mrs. G's pulse returned, and she started breathing on her own.

From the Brigham tower, facing north, the view from the tenth floor runs down along Brookline Avenue toward Fenway Park and the famous CITGO sign less than a mile away. During the 1970s and 80s, Dwight Evans patrolled the outfield at Fenway. To most Red Sox fans, Evans was the greatest right fielder of all time, and much of that regard is due to a play he made in Game 6 of the 1975 World Series. In Boston lore, it has come to be known simply as "The Catch."

It was the top of the eleventh, one out, Ken Griffey on first, with Joe Morgan stepping to the plate. Evans remembers thinking through everything that might happen. "Good fielders do that," he said later. "That's how great plays are made." Morgan hit a long fly ball to right, a potential disaster. As Evans recalled it, "I turned towards the line because normally a ball turns toward the line too. Well, this particular ball didn't turn. This particular ball stayed straight, and you ask any player, when you lose a ball, that's a scary situation. No one was more surprised than me. I jumped, and my glove went behind my head, and the ball landed in my glove. I'm glad, thank God, I caught the ball, because if not, it goes in the stands. To me it was the most important catch I'd made. It wasn't the best catch, it was an awkward catch, but there was a reason it was so awkward, because I did lose the ball for that split second."

After the catch, Evans's throw to first base was twenty feet wide to the left, requiring an alert play by the shortstop and the first baseman to double up the base runner. No one remembers the errors Evans made that season, or the strikeouts. They remember the great moments. Very few people get a chance to make a catch like that.

Would Mrs. G have died had I gone home early? During any code there is a laundry list. Most codes deal with catastrophic but common medical problems like heart attacks, pulmonary embolus, and aortic rupture. The code team would have gone through the list in order to address those possibilities. The code leader would eventually have realized that it was none of the above, that it was a neurological problem. Hannah knew that. She would have connected the dots. Whether she would have had the middle-management ability to pull everything together by ordering people around and saying, "Get the neurosurgeon up here, get the mannitol in," that would have been a test, but she would have done it, she would have made the catch. All I did was bridge the gap, maybe speed it along.

The next morning at about 11:00, I went down to the ICU to gauge the aftermath of what I thought might be a terrible situation. Elliott was there.

"I hear you missed the game," he said. "Too bad. They pulled it out in extra innings."

"So did we," I replied.

Mrs. G had developed some complications from the CPR, notably a sharp pain in her chest with every breath. She had no idea how her ribs got cracked. But there she was, sitting in an easy chair eating breakfast, with a ventricular drain coming out of her head and snaking down an IV pole to a reservoir. She was wide-awake, and I asked her if she knew who I was. She said, "No, I've never seen you before."

"Well," I replied, "I've seen *you* before. And it's very nice to see you again."

3

The State of Confusion

Two characters in search of a neurologist

Within a span of five years Gordon Steever's life had been reduced to a half-mile radius of a bowling alley in Dorchester, Massachusetts, where, until recently, he had worked as the day manager. He lived five blocks south of the lanes, his ex-wife lived four blocks north, he could walk to the superette three blocks east or to his favorite diner just down the street, and on rare occasions, if absolutely necessary, to the Fields Corner subway station. At the age of sixty-seven, his routine kept him within a purlieus that had enclosed almost the entirety of his life. In other words, Gordon was a townie. He didn't own a car. He lived alone in a two-room flat above a beauty salon. He still owned an analog television. He listened to the Celtics and Red Sox on a Zenith transistor radio. He went to mass only on Easter and Christmas Eve, but still avoided meat on Fridays. He rarely saw his grown children.

On the evening of May 26, Gordon was feeling increasingly agitated. Earlier in the day he had tried and failed to fill out the paperwork required to collect unemployment checks. He had been let go from his job at the bowling alley two weeks earlier, not so much for erratic behavior, something the owner could live with, but for not showing up, which the owner would not tolerate. As his mood darkened, Gordon walked through streets that he had known all of his life, streets that suddenly seemed oddly unfamiliar. Finding himself on Gibson Street, without recognizing it for the first time in sixty years, he walked through the front door of the Boston Police Department District C-11 Headquarters, and started haranguing the desk sergeant.

"I can't take it anymore," he said.

"Can't take what?"

"A lot of the kids there, a lot of the kids. A lot of people, not around here now, but not like that one."

"What's your name, fella?"

"Gordon."

"Where do you live?"

"Dorchester."

"Okay, Gordon, what's your address?"

"I can't remember what the address was."

"How about *your* address, where your house is?"

"Well they kind of changed a little bit, because they've . . . they're not quite as good as . . . by next week I think it'd be really perfect, people would be able to understand what these things are."

"What things?"

"You don't have no idea, you haven't even thought of it, or anything else. Every person in this world, unless they have had health problems, they're done. It's true! You should come down here. You sit there and look. The intelligent person would be a person just sitting there and looking. Just like that."

To the sergeant's trained eye Gordon did not look drunk. Nor did he look sick or delirious or drugged. What he did look like was a very confused, wiry guy in his late sixties, bowlegged, with a bristly mustache, short gray hair on a balding pate covered with peeling skin. Maybe he'd been a seaman, not Navy, more likely Merchant Marine.

The guy had seen better days, and now appeared to be consumed by resentment over some perceived slight. He wasn't threatening, but, as the sergeant later told the EMT, "He's either got a hair up his ass or something's really wrong with his head. Most likely both."

After the EMTs dropped him at the nearest hospital, Gordon became paranoid and agitated, so much so that they could not get him into the MRI machine. Another ambulance then took him to the psychiatric ward of the Faulkner Hospital, where it was determined that Gordon's problem might not be psychiatric. They wanted him checked out first by a neurologist. So Gordon got to take one more ride.

———

Three weeks later, on a cloudless June day just after lunchtime, Walter Maskart was driving his wife to a doctor's appointment when he missed his exit off a traffic rotary in Braintree, and ended up heading east instead of west. Had he taken the exit MapQuest had directed him to, he would have arrived at the medical associates in plenty of time. The office building was located just off the rotary. Instead, Wally drove through East Braintree, then South Weymouth, and eventually down to Bridgewater before getting his bearings and heading back north on Route 3. He had left his house at 1:00 for a 2:00 appointment, allowing himself plenty of time. He did not arrive until 4:00 p.m., and then got it into his head that he had been driving for five hours.

The receptionist told Wally that he would have to come back the next day, but he insisted that she go explain to the doctor what had just happened, that he had lost track of things, that something was wrong with him. The doctor, an oncologist who was treating Wally's wife for lung cancer, came out and talked briefly to Wally in order to see whether he was oriented: did he know who he was, where he was, what day it was, did he know how to get home? Satisfied that Wally was indeed "oriented times three," he told him to return the next day. As the couple walked back to the car, Wally's wife started to cry, not for herself, but for her husband.

At the age of seventy-eight, Wally had reached a breaking point. His wife's chances of beating the cancer were merely fair, his daughter's marriage was falling apart, and his business—a party supply store—had just failed. He now had to run a household, care for his wife, cook the meals, manage his stock portfolio, all while dealing with his own health issues: diabetes, hypertension, COPD, obstructive sleep apnea, and congestive heart failure. The last straw that sent him to a local hospital, not for his wife but for himself, was the confusion. After a complete medical examination revealed the respiratory and cardiac problems, it was decided that Wally would need the resources of a larger facility. In addition to his confusion, it seemed, Wally would occasionally stare into space and blink in clusters. It was thought that he might be having small seizures, so he was sent to us, and we fitted him with a twenty-four-hour EEG monitor.

Wally Maskart was placed in a room next to Gordon Steever, two doors down from Mrs. Gyftopoulos. Wally and Gordon, although in the throes of acute confusional states, were a study in contrasts. Wally engaged the world, while Gordon hid from it. Wally witnessed Mrs. Gyftopoulos's code blue and wrote about it in his journal, while Gordon kept to his room and stared at the floor. They had a few things in common. The first, which was speculative, was that Gordon had been Wally's daughter's seventh-grade basketball coach. The second, which was unfortunate, was that both of their spouses were battling cancer. The third, which was definite, was that I had no idea what was wrong with either one of them. The fourth, which was obvious to even the most casual observer, was that something was very, very wrong with both of them. We couldn't rule out the possibility that their confusion was not a neurological phenomenon, but a psychiatric one. We also couldn't rule out the possibility that in both cases it was life-threatening.

Callie, the second-year resident with the idiosyncratic word selection, presented Wally's workup in the conference room on the morning of Mrs. G's code: "Mr. Maskart is a seventy-eight-year-old right-handed man who was recently admitted for episodes of confusion. He has a history of coronary disease, the usual slew of hypertension, hypo-adrenal whatever, diabetes, obstructive sleep apnea. He's presenting again after persistent episodes of confusion: just wacko, per the family. He's been getting lost while driving, he's been having episodes of what he describes as déjà vu. His son says he seems to be a little more confused in the mornings. He seems at times to be unable to work the television remote. He's really into toy trains . . ."

"Model railroading," I corrected.

". . . whatever, but it seems that recently he bought the wrong kind of train things, which his family says is just not like him."

"What things?"

Callie looked through her notes. "A tinplate locomotive, whatever that is."

"Continue."

"He was admitted to medicine [the medical ward] on the twenty-fifth. He had three days of continued déjà vu, increased sleepiness, and increased confusion. Then he was readmitted and sent here. The other thing that's complicated this is that he seems to be playing around with his medications at home."

"What do you mean?"

"He has psoriatic arthritis and is on prednisone. He seems to be getting his meds confused. Apparently he likes to tinker with them, and he might be taking his wife's medications. She has cancer."

"Has he had a tox screen?"

"I truly hope so."

The presentation took longer than most because the history was clogged with a full battery of tests. To use another of Callie's favorite expressions: Wally had been worked up "up the wazoo." Many of the results were still pending, and some would take weeks to come back. In the meantime, I suggested getting even more esoteric and expensive tests. As Elliott never tires of telling me: "Hey, it's other people's money. Don't start believing it's our responsibility to reduce health-care costs." To which he was quick to add: "Not for quotation, by the way. I can just see *that* on the front page of the *Globe*."

It may sound trite to say that *confusion* is the most confusing syndrome in medicine, but it is. A confused person behaves in a way so foreign to common experience that it can be unnerving, even for professionals. It is an alternate state of being. Portrayals of confusion in popular culture—the town drunk, for example—may look funny, but in the case of a truly confused person, the sight of someone who can't find his own mind can be overwhelming.

As a neurological entity, confusion is up for grabs. Any young researcher hoping to make a name for herself might consider starting a large-scale study of it. Or not. It could easily eat up a budding career. An understanding of confusion has yet to be operationalized in the way that stroke or neuropathy or Parkinson's disease have been. It is not, technically, a disease, but a syndrome, a collection of problems. Clinically, confusion is defined as a loss of the usual clarity, coherence, and speed of thinking, but this description, while accurate as far as it goes, captures only a snapshot of confused behavior. In some patients, blood tests will reveal a metabolic cause that can be addressed: low levels of blood sugar, for example, or high levels of carbon dioxide. But as with many neurological conditions, there are no definitive tests for confusion. We have to rely on the clinical exam, the patient's history, the story as given by family or neighbors, and any little clues we can unearth. A tinplate locomotive, for example.

On my first visit with Wally Maskart, he seemed fairly lucid, if not completely with it. After introducing myself, I got right down to business.

"You're a model train guy, right? HO gauge?"

"No. O gauge."

"Good. I've got to come see your setup someday. I have N scale and HO. A lot of scale buildings, too."

"So do I."

Wally was sitting up in the hospital bed. His face was ruddy, his gray hair plastered down and tousled by the recently removed EEG leads. He was a hunched, flushed, somewhat pudgy fellow, constantly short of breath, and thus incapable of speaking with ease or completing a sentence without pausing to inhale some of the oxygen coursing through the tubes under his nose. He had the appearance of a guy who was once physically active, if not robust, and was now deflated.

"What brought you to the hospital?" Hannah asked him, and he proceeded to tell the story of getting lost for five hours.

"So that's where I screwed up, getting all confused again," he concluded.

"This driving for five hours, you'd never done that before?"

"No. I knew where the place was. I'd been there before."

"So where *did* you go?" she asked.

"I can tell you exactly what happened. I looked up MapQuest. On MapQuest there's a big circle, a rotary, and you enter the circle, and MapQuest says bear right, and then bear right again. But what I actually did is take the *second* right and *then* bear right. So I missed the building, which was right there, and got three towns away."

"Three towns? And that's the only time that's happened?" Hannah asked.

"Yeah."

"You're not confused now," I said. "You're very clear."

"I am now. I told my wife, don't worry, but I was three towns away. And I made it home alright without any problem at all. The only problem was when I told her let's go home, she got teary eyed."

"But that was three weeks ago. What made you come back to the hospital?" Hannah continued.

"I got fuzzy thinking. I was thinking fuzzy."

It wasn't clear what he meant by that. As cogent as his answers seemed, two things about Wally were definitely off. One, we soon discovered, was that he would drop off to sleep for short periods of ten or fifteen seconds in the middle of a conversation. The first time it happened, I was concerned that it might have been a seizure.

"Are you with us now?" I said, trying to assess his orientation.

"Yes."

"I thought you might have just had a little seizure."

"No, I fall asleep."

"You just fell asleep right now?"

I had him count backward from one hundred to see whether it might elicit a blank moment that signaled a seizure, but it didn't. He made it to zero without an error.

"Are you a talkative guy by nature, Wally?"

"No," he answered, then thought about it. "Let's put it this way . . ." Then he paused, and smiled. His neighbor on the other side of the curtain yelled out: "HE GETS STUCK ON THE RED SOX!" That, as it turned out, was Wally's second problem.

"I guess I am talkative," he said sheepishly. "I guess I talk a lot."

This would become evident over the next few days. He routinely parked his wheelchair at the door to his room and flagged me down whenever I walked by, convinced not only that I was his neurologist, but that I was at the same time a heart specialist named Sanjay Sanjanista, and that I had performed the miraculous surgery that had brought his wife back from the brink of death several years earlier.

Wally Maskart had arrived on the ward in the depths of a confusional state that frustrated all of our attempts at diagnosis. He was depressed, sometimes manic, never at a loss for ideas, but frequently at a loss (due to his COPD) for the breath with which to voice them. Upon meeting him, I recognized that the demons besieging him were so deeply entrenched that we might never arrive at a diagnosis. To pass the time, I assigned Gil, one of the medical students on the team, the task of tracking down Dr. Sanjanista. He came back and informed me that no such person existed, and that his best guess was that Sanjay was a chimera inspired by a memorable if not particularly talented contestant on *American Idol* named Sanjaya. Despite his vocal flail-

ings, Gil informed me, Sanjaya's looks had earned him a cultish female following known as Sanjanistas, whose mission in life was to
stuff the online ballot box in order to overextend Sanjaya's stay on the
show. As if that weren't enough, Wally also got it into his head that I
had scored the winning touchdown for Boston University in a memorable game against Northeastern in 1962. "I was there!" he told me.
Alas, I wasn't.

Wally's insistence on his personal connection with me could have
been discounted out of hand were it not for his precision on other
points, such as his ability to recite the decimal expansion of e (the base
of the natural logarithm) to twenty decimal places. He even wrote it
down, and I confirmed that he was indeed correct. He could also dia-
gram the chemical composition of acetone, and seemed well versed on
the career of the Red Sox right fielder Dwight Evans, except for his
claim that Evans had just died. Evans, Wally insisted, had been cal-
lously traded to the Mets near the end of his career, an assertion that
Gil was able to refute via a Google search, although Wally wouldn't
accept it. He also claimed he had been at Game 6 of the 1975 World
Series, and witnessed Evans make "The Catch." At that moment, he
had turned to his son and said, "Tommy, sit down and relax. God's on
our team's side. There's no way they're going to lose this game." And
he was right. Carlton Fisk won it for the Sox with an historic home run
in the twelfth.

Wally also told me that he had his eye on a prewar tinplate locomotive that I knew cost around $2,000.

"You didn't really buy it, did you?"

"I don't remember. I mean no, my wife would never let me."

Ironically, the "key to this whole thing," to use Wally's phrase,
was staring me right in the face. He had indeed bought the locomotive, a Lionel Red Comet, a sweet little piece of machinery that was
well beyond his budget. (I know this because I had had my eye on
the same locomotive, and my wife would go ballistic if I ever bought
one.)

Marsel Mesulam, the prominent and prolific Northwestern University neurologist, has taken the lead in trying to define confusion, and has focused on what he calls "the attentional matrix." The matrix, as he views it, is a place that serves as a temporary register of items that make up links in a sequential thought process. Confusion, according to Mesulam, disrupts tasks that require attention, and by inference, it represents a disruption within the matrix.

I think that's far too limited. When you sit at the bedside of confused patients, this is not what you see. They are inattentive, to be sure, and you can use that as an identifying feature, but many other things enter into it. For example, misperception and illusion (one thing being misinterpreted as another), topographic and spatial difficulty (they can't draw, they can't copy boxes), a loss of temporal coherence (difficulty connecting one moment to the next), and a hard-to-define language trouble. Those are all elements of various confusional states, and they do not derive from inattention so much as accompany it.

Disorientation, for example, a staple of confusion, is not simply a result of inattention. Confused people have little or no insight into their predicament, and they cannot absorb their circumstances when I tell them that their minds are not working. They get lost and, most distressingly, become irascible, agitated, wildly unmanageable, or in some cases, their minds collapse inward, and they become silent and akinetic, as though all of the thin threads of cognitive life that bind thinking into a flowing stream have been severed. There are more interesting, restricted cases of confusion, such as patients who cannot recognize their own paralyzed limb (a condition known as anosagnosia), but unlike the confusional states manifested in Wally and Gordon, these forms of confusion result from structural damage in specific brain regions. Yet each of them in mild form is part of the global confusional syndrome of fuzzy thinking as the sum of a hundred little failed areas of the brain, and the disruption of the specialized tasks performed in these areas.

My dilemma? Wally and Gordon did not fit these patterns or fall into neat categories. Neither man was particularly inattentive. Both were aware on some level that they had a problem. In each case, whatever the problem was, it had to be at a submicroscopic level, too small to show up on a CT or MRI scan. The fact that electrical activity between nerve cells is highly disordered in most confusional states, however, does show up in the slowing of the EEG waves. This was the pattern we found in Gordon Steever's tests. It did not tell us *what* was wrong with him. The tests merely confirmed that something *was* wrong with him.

Gordon Steever had been living on the ward in a room next to Wally's for two weeks by the time I arrived on the scene. Elliott had admitted him, had even spoken to the police sergeant who had sent Gordon to the hospital, and was as baffled as I was. Gordon would remain on the ward for two months after my service ended, mostly on the taxpayer's dime, with the hospital picking up the rest of the tab. Although he was gruff, profane, unpredictable, and given to angry outbursts, he was well liked by every member of the staff. He had the voice of Cliff Clavin from *Cheers* and the manner of Louis De Palma from *Taxi*, but unlike Louis, Gordon had no artifice. He was neither manipulative nor conniving. He steadfastly avoided eye contact. At first I thought he might have been suffering from Creutzfeldt-Jakob disease, known to us as CJD, and more popularly as the human form of mad cow disease. Because of the rapid decline of his mental state, nothing else seemed to fit. Despite his inaccessibility, his sheer irascibility, I instinctively liked him, and looked forward to our visits. Even on his worst days, I would drop by simply to take a mental break, and I would usually drag a few members of my team with me.

"HELLO!" I yelled in order to be heard over the air handler. "How are you, Gord? It's great to see you! How ya doin'?"

Hannah and young Gilbert, the medical student, lingered noncommittally just inside the threshold, as if at the edge of a precipice.

"Good," Gordon muttered almost inaudibly.

"Are you confused?"

"Kinda. Everybody's always confused."

"Am I confused?"

He paused. "Nope."

"Mr. Steever, do you know where you are?"

"Somewhere in this world."

Gordon spoke with a flat affect, a tone of resigned indifference rather than one of confusion. He sat in the easy chair between the bed and the window, wearing hospital-issue blue drawstring pants and no shirt, perpetually bent forward like a ballplayer in the losing team's dugout, alternately anxious and agitated. His nurse, Carmen, stood by the bed, poised to intervene if necessary.

"And what kind of place is this, Gordon?" I asked. "Is it a supermarket?"

"I always get mixed up," he said quietly. "I'm in a box, and it's the wrong box, and I'm just in another box."

He was right. Gordon was one of the few patients on the ward with his own private box—a single room with a view. Through its windows he could look down onto Francis Street and see the gleaming metal-and-glass facade of the new Shapiro Cardiovascular Center, and beyond that the Riverway and Brookline Village. It was a beautiful afternoon, with the sun dipping just low enough to produce some contrast in the tree canopy that welled up beyond the nearby rooftops. But Gordon didn't look out the window. Most patients don't. No matter how spectacular the weather, no matter how striking the view, no one seems to notice, including the nurses, and especially the residents. As far as they're concerned, we could be anywhere.

"Do you know what city we're in, Gordon?" I asked.

"City where people work and live."

This was the common theme. Gordon responded tangentially. If he were merely demented, he would have said something like "New Jersey." His was a unique type of confusion, with speech that approached what my professor, C. Miller Fisher, used to call *amphigory*, or nonsense speech.

"He looks a little drugged to me," I said to the nurse.

"He actually hasn't had his Haldol for eight, ten hours, because his IV came out."

"The key," Gordon interjected, staring blankly at the floor, "is when you ever feel good bowling, you can stand back at the line, and you can say, 'Look now, count one-two-three.'"

"That's the key?"

"That's the key."

"Are you hallucinating, Gordon? Are you seeing things right now?"

"No."

"Good. What's your address?"

"My address? Let's see. There's a dog in here talking, so he can just say, 'Hey, do you want to know the best days? Look here!'"

"How the hell does the brain assemble *that*?" I asked Hannah, somewhat rhetorically. "It's certainly not just aphasia. His diction and tone are perfect and his comprehension near perfect. It's his internal conversation that's muddled." She responded by puffing one cheek.

Apparently fine just a month ago, Gordon had rapidly degenerated into this state. After two weeks of tests and bedside exams, none of my colleagues could say what was wrong with him or what had caused this. My initial thought was that these strange responses were rattling around in his temporal lobe, but could not connect with each other.

"He's talking nonsense," I said, "but there's no brain module that creates nonsense. The question is: Could this be the result of the degradation or absence of some other function? Is it a problem with the connections between brain modules?"

"My theory on all of this will straighten it all out for all of you," Gordon interrupted. "You want to know why?"

"Please."

"Because none of you care. And if you *do* care, it's going to go to the other place. And you can move this way or that way, or upstairs, and then you'll see, probably, a dog. Probably got a nice little dog, a smart dog. It's intelligent enough. Mostly, the dog will curl up for you."

Gordon Steever's mind flitted like a waterbug across the surface of his life's experiences. There was no depth. His speech lacked any internal logic. Not even schizophrenics talk like that. It was almost as though we were looking in on a performance.

"Are you happy?" I asked.

"Pardon me?"

"Are you feeling happy now?"

"Ah! I would be happy if I'd seen it."

"Ho, boy!" I said to Hannah. "You can't argue with that. He's like a philosopher, but he's talking Martian. I don't think there's any emotional valence under there. When I ask him something, he responds. It's not a waking dream, it's usually nonsense, but he keeps returning the ball. How does he do that?"

I turned back to him and said, "Gordon, let me ask you something: Is it warmer in the summer or the South?"

"The South? It's hot."

"Okay, good. Then tell me, Gordon, do helicopters eat their young?"

"It is true," he replied blankly, with no indication that he had processed the question at all.

Gilbert the medical student cocked an eyebrow. "Helicopters?"

"It's almost diagnostic," I explained. "I'm just talking to the confused brain on its own terms. If an aphasic patient gets the joke, it probably means he's faking."

"So Mr. Steever's not faking?"

"No such luck."

"Instead of saying good luck," Gordon suggested, "you say good ruck. Think about that!"

"Okay, I will." I turned to Hannah. "Does he have alcohol in his history?"

"His wife says no," she replied.

"Are we sure this has never happened before?"

"Per his wife he was managing a bowling alley."

"It doesn't have the mad cow feeling to me at all. It's some kind of encephalopathy. Has he been poisoned? Is he in status? We decided he wasn't, right?"

"Correct," Hannah said.

Status is short for *status epilepticus,* a state in which a patient is constantly seizing. In this case, without physical convulsions, it would properly be called nonconvulsive status epilepticus. But Gordon wasn't seizing. The Haldol, an antipsychotic, had made him worse, suggesting that he was also not psychotic.

In the course of our inquiries, we would discover that Gordon had three grown children, possibly stepchildren, none of whom visited him. His wife, either estranged or ex, was undergoing chemotherapy, and also did not visit, although we did manage to reach her by phone. In time, Gordon would become a ward of the hospital in order to facilitate medical decisions he was unable to make on his own.

"Gotta look for good ruck, good ruck, good ruck to all," Gordon said to no one in particular, and added, "I can remember coming down in college. I turned out to be sitting there talking and so forth and so on. Oh, yeah, he was a hippie or a bippy. What's the difference? It's not going to make any difference, so you can turn right around." And with a wave of his hand, Gordon gestured everyone out of the room. But to his consternation, we stayed.

"Mr. Steever," I said in a last stab at a clinical exam, "can you show us two fingers?"

And he gave me the finger with both hands.

"Right! He's got good comprehension."

Saying "Go fuck yourself," as Gordon did to me on several occasions, was not inattentive. It was ornery, irascible, and rude, but not entirely irrational, and certainly not inattentive. Gordon could be extremely focused on the details of our comings and goings in his room. Our presence pissed him off, and he let us know it, yet day after day he could not tell us his address, how many children he had, or what kinds of jobs he'd held. And not out of obstinacy, because he was fine with other details. He was disoriented and did not know where he was or why he was here. Gilbert the medical student recorded this as "orientation times one," again that unfortunate and unenlightening phrase. But Gordon also had his moments of lucidity. On good days he could tell us where he went to college and the name of the bowling alley where he had worked. He could, on occasion, be quite congenial.

During the three months of his stay at the Brigham, we performed every conceivable test on Gordon Steever, and they all came up either negative or inconclusive. He became very popular with the staff, less profane, but no more oriented than on the day he arrived. When he finally ventured out of what Hannah referred to as his man cave, he enjoyed tossing a tennis ball across the nurses' station to anyone who seemed willing to field it. Eventually, he was transferred to a psychiatric nursing home, and we lost touch with him for six months. He was a ward of our hospital, and the best we could do was fob him off on a place where he would languish, and quite possibly die of neglect. He was gone but not forgotten. It was not clear where his confusion came from, but for now we knew where it was going to take him: Saugus, the only town in which Hannah could find a facility that would offer a bed.

In the room next door, Wally Maskart sat awake through most of the night, as he did night after night, frantic. In his sleep-deprived,

hyperactive state, he would write for hours on end, collecting his dis-
jointed thoughts and memories in a loose-leaf notebook, producing up
to twenty pages a night, and each morning, Hannah would photocopy
some of the more interesting pages and insert them into Wally's medical notes. Hannah was searching for clues. One night she found one.

While most of his jottings were gibberish, Wally had moments of
lucidity during which he displayed a practiced prose style. In one burst
of creativity, he wrote a brief memoir of the summer of his twelfth
year, when his parents sent him to stay with his uncle, a fisherman in
Nova Scotia. He didn't understand how or why an East Boston kid like
himself wound up on the open sea, hauling halibut into a dory, but he
reveled in every day of it. The sight of baleen whales surfacing every
morning beyond the bay, he was told, was the result of their being
driven closer to shore by the German U-boats that infested the waters.
It was 1943. The summer culminated in a village tradition, in which
the fishermen and their families came together on the water to haul in
a large co-op net trap, some hundred meters across, containing a large
school of pollock. As the depth of the massive net was drawn to within
six feet of the surface, the water came alive with the thrashing of thirty-
to forty-inch-long fish, and the fishermen and women scooped them
into the dories using pole nets. In the process, fish scales flew like sleet,
covering everyone from head to foot, leaving only the whites of their
eyes showing. Wally recalled being assigned a spot next to a young girl
wearing a yellow slicker that gradually turned silver under a spray of
glistening scales. Hannah decided that this was the most serene moment in Wally's long life: standing next to a cute girl, being waist-deep
in thrashing fish, living in a safe place among strong, stoic, caring
people, covered with fish scales. It was probably the last time that Wally
was at peace with the world and with himself.

When she showed me the journal, Hannah said, "It reminded
me of something Gordon said the other day. I don't know why it
stuck with me. He said, 'Do you want to know the best days? Look
here!'"

That's when it hit me: graphomania (the obsessive writing), alliteration (the Sanjay Sanjanista business), and the biggest red flag of all: the tinplate locomotive. Only a wealthy collector buys a $2,000 locomotive, and then only to put on display, not to run it on his setup.

"Did we get a family history for Wally?" I asked Hannah.

"More or less. We talked to the son."

"Which one?"

"He has more than one?"

"Yes."

Wally did indeed have two sons. The first one who visited had provided the family history that the residents had relied on: no history of psychoses, no similar events in the past. His confusion had come out of the blue. Except that it hadn't. We had talked to the wrong son. The other son, Wally's *bipolar* son, who showed up a week later, painted a very different picture, one in which a history of psychosis ran deep in the family, deep in his own past, and deep in his father's past. Wally had snapped under the pressure. He'd had a nervous breakdown, something, as we would discover, that had happened before. That's when we knew he would probably be okay.

Psychosis is a special type of confusion with its own reality, an internal reality that is consistent only with itself. It does feature a connected, ever-flowing "stream of thought," to use William James's term (not a stream of *consciousness*, a phrase that James came to dislike). In his *Principles of Psychology*, James claimed that all of us carry on a virtually continuous internal conversation. While a psychotic's internal stream may seem bizarre and disconnected, it has its own internal logic. Anyone could follow it, according to James, if they were standing in the waters of the stream. Wally Maskart's thinking might have been crazy (in fact it *was* crazy), but it wasn't stupid. My job was to step over the boulders and out into the middle of the current, and try to lead him back to shore.

That wasn't possible with Gordon Steever. Gordon was severely confused, but not psychotic. During his stay in the hospital, and increasingly during the weeks leading up to it, he lived in a fog, in a miasma in which there was no connection between one moment and the next, no continuous stream of thought, no path of return. Wally, on the other hand, was caught up in a very different struggle. He was delusional but he could connect. Even though he insisted that I was both Dr. Ropper and Dr. Sanjanista, he was not detached from all reality. A confusion of this type is in most cases a reversible state because it is a reflection of the dynamic functioning of nerve cells. If the causes are addressed, the patient will get better. It just takes time. If the causes are not addressed, the confusion takes over.

When I was a boy I had a Jimmy Piersall baseball glove. Piersall was my favorite ballplayer, and he was famously nutty. He made his major league debut with the Boston Red Sox in September 1950. Before he stepped up to the plate, he turned his bat upside down, and with the knob end made an X in the back right-hand corner of the batter's box, something he would continue to do throughout his career. Wally Maskart saw Piersall play right field at Fenway Park many times during the '50s, and watched with passing interest as Piersall's life fell apart due to a bipolar disorder.

Piersall was a local boy from Waterbury, Connecticut, and he became one of the best fielders in the game, but he was plagued by a past that included a deeply troubled mother who had been institutionalized, and a demanding father. Frequently involved in brawls, bizarre on-field stunts, and tantrums, Piersall racked up a score of ejections over the years. In his autobiography, *Fear Strikes Out*, he tells of being admitted to a psychiatric hospital in 1952 after a nervous breakdown. "I ran away," he recalled. "I had just gotten so wound up that I lost all control of my memory." Treated with lithium, he would last seventeen years as a player, and many more as an instructor and broadcaster. As he once admitted, "Probably the best thing that ever happened to me was going nuts. Who ever heard of Jimmy Piersall before that happened?" No baseball fan would ever forget him after he hit his hundredth home run at the Polo Grounds in 1963, and ran the bases facing backward.

Although I would never again see Gordon Steever once he left the Brigham, I did visit Wally Maskart three weeks later at the psychiatric hospital where Hannah had managed to place him.

"I feel much better," he told me.

"Are you oriented? Is your mind clear?"

"What used to take ten minutes now takes a few minutes."

"Are you writing in here?"

"Yeah, I'm writing my life history."

Among his topics were accounts of Mrs. G's code blue, a summer cottage he had bought in Mattapoisett thirty years ago, and his time working for Polaroid in the 1970s. He looked much better, sounded much better, his memory was improving.

"I feel like I could go home today," he said.

"Wally, who is Sanjay Sanjanista?"

"You are."

"How does that work? How can I be Dr. Sanjay Sanjanista and Dr. Allan Ropper at the same time?"

"That's your first name and your second, and you're my wife's cardiologist, and that's when I first met you, and in your office there's a sign, 'Go Bruins,' and that's what attracted my eye, and that's how it got started."

"How are you sleeping?"

"I never did sleep well. I have to get up and pee a lot. Then there are these damn alarms here. But I feel like I'm much more rested."

"Will you be able to manage at home?"

"Oh, yeah. The house is all set up."

There was one other issue that bothered me. "Do you remember talking about Dwight Evans? You were under the impression that he had died. You might have been confusing him with Dick Williams, the Red Sox manager, who did die a few weeks ago."

"You know what? I think I said . . . no . . . here's what happened. The greatest right fielder the Sox ever had was traded, and what reminded me of it was when the coach died. Yeah, he was the coach when they won the World Series. And they had the greatest right fielder ever. He's the one that ran the bases backward."

"That sounds like Jimmy Piersall. Different right fielder, but a great one."

"Oh, yeah, was it? No, you're right, it was Jimmy Piersall."

A light went on. It *was* Jimmy Piersall, and *not* Dwight Evans, who had been traded by the Red Sox, who had hit his hundredth home run while playing for Casey Stengel, who was traded soon after, but lasted four more years in the majors. I had waded into Wally's stream of thought, and had met him somewhere in the middle. If Piersall could make it, I thought, so could Wally. And he did. With the help of Seroquel and lithium, he would soon be discharged and return to the problems that had driven him to us in the first place.

"I sold the locomotive," he said. "I took a loss."

"Better to be rid of it," I told him. "You can't pull freight with it anyway."

4

My Man Godfrey

A poor sort of memory that only
works backwards

In the late 1970s I had a patient, a salesman who drove from Philadelphia to Boston unwittingly, and made it as far as Leverett Circle, a traffic rotary near Massachusetts General Hospital, where he got stuck driving around and around for almost an hour. Eventually a policeman noticed him, pulled him over, and said, "Is everything okay?"

The man replied, "I don't know how I got here."

The policeman had the good sense to send him to the emergency room, where he was examined by a junior resident who found nothing amiss beyond the memory loss. Concluding that it was an episode of transient global amnesia, or TGA, a dramatic but entirely benign condition, the resident came to me to approve a discharge in anticipation of the expected return of the patient's memory. Although it may sound serious, transient global amnesia can be set off in a variety of ways, has no obvious cause, and will usually resolve within hours, leaving no permanent damage.

At that time, I was the senior resident on the neurology service at

Mass General. After examining the man (I'll call him Godfrey), I recall saying, "How can a brain function at such a high level and have no memory?" Godfrey had driven all the way from Philadelphia, yet he remembered little of the drive. He did remember getting into the car twelve hours earlier, but had trouble remembering the meal he had eaten five minutes ago.

Godfrey was in his midfifties. He was a short, somewhat plump man, with a double chin and two-tone eyeglasses that were out of date at that time although, ironically, back in style today. He was extremely pleasant, and despite, or perhaps because of, his confusion, he didn't mind hanging around the hospital and getting a little attention.

As I moved around the cramped emergency room cubicle to examine him, I repeatedly bumped backsides with the resident working in the next bay (the semicircular curtains at Mass General had a restrictive diameter). I sensed that there was more to this case than met the junior resident's eye. Godfrey couldn't retain names—mine, the resident's, the name of the hospital—and could not believe where he was. "Jesus Christ, Boston? You're kidding!" As he remembered it, he had set out for Harrisburg, Pennsylvania, on a sales call. After a while, he came to accept that he was in Boston—he managed to retain that much—and even recalled that he had eaten pancakes for lunch, but when questioned later, he drew a complete blank.

"If this isn't a transient amnesia," I asked myself, "what else could it be?" There were only a few possibilities: a concussion, a viral infection in his brain, ongoing seizures, or a stroke. All four affect memory. Three out of four are life-threatening. His CAT scan was normal, seizures seemed unlikely, it wasn't drug-induced, so it had to be either transient global amnesia or a stroke.

At our first encounter, Godfrey's memory for the days, weeks, months, and years before this event seemed intact, as far as I could tell without being able to verify the name of his third grade teacher or his high school sweetheart. He reported a vague sense that something peculiar was going on with his mind, but he was not alarmed.

"The hospital meals aren't bad," he told me. "Those pancakes were spectacular."

"What are you talking about?" I said. "This food is terrible. It's what *we* have to eat every day." But twenty minutes later, when I mentioned the meal again, he was perplexed. He had lost all memory of the pancakes. If it really were a transient condition, a benign TGA that would resolve in a few hours, there was no way I could justify admitting him as an inpatient. Yet I didn't want to send him on his way just yet. Something didn't feel right. His memory had holes in it and the problem was lasting too long. Without an alternative diagnosis, my only recourse was to admit him to the overnight ward.

The overnight ward was a way station where we could keep an eye on patients for a while without officially admitting them. It consisted of eight beds in the back of the Emergency Department, and it was run by the residents. We senior residents could admit people there, not to the hospital so much as to ourselves, for up to twenty-four hours. It was a great invention. All the drunks going through withdrawal, neurotics (to use an outdated but perfect term) we didn't want to send home but didn't want to admit, other characters we weren't sure about—we could give them a bed and observe them. In the modern hospital environment, an attending physician would dress down a resident for admitting a patient like this one, but something seemed fishy enough about Godfrey that I decided to keep an eye on him overnight. The next day, twenty minutes before he was due to be discharged, he took a turn that made me awfully glad I had.

I found myself recounting this story recently during morning report, a fixture on every resident's daily schedule. Each weekday we gather in the library at 7:30 a.m., where two senior residents present cases they are about to discharge from our Emergency Department. On that particular morning Hannah chose to present the case of a 62-year-old Colombian woman who had developed memory problems after a weekly ritual involving sex with a casual acquaintance. After the tryst, a neighbor brought the woman to the hospital because she seemed very confused. In the emergency room she kept repeating the same questions over and over at thirty-second intervals: "I feel fine." "How did I get here?"

"I asked what the problem was," Hannah said, "and that's exactly what she said: 'I feel fine. How did I get here?' So I explained that her friend brought her in because she seemed to be behaving oddly, and she thanked me. All of the social graces were there: voice modulation, natural body posture, alertness, eye contact. Everything seemed normal until thirty seconds later when she said it again: 'I feel fine. How did I get here?'

"I told her that her friend thought something was wrong," Hannah continued. "I asked her whether she had a sense that she was confused, and she said, 'You know, I feel a little odd, but I think I'm okay.' So I checked her orientation: What's your name? Where do you live? Who is the president? How many fingers am I holding up? After answering all of these questions perfectly and my explaining why her neighbor brought her to us, she looked at me pleasantly and said, 'You know, I feel fine but how did I get here?'"

This time the line got a laugh.

The gravity of memory problems is often disguised by their risibility. Someone in the throes of aphasia or agnosia, that is, someone whose perception functions properly, but whose processing does not, can unintentionally crack up a room full of trained specialists. Cases like these, replete with malapropisms and verbal absurdities, are more bizarre than scary. Transient amnesias in particular may last a few hours, almost always less than a day. Any number of things can trigger an episode, or nothing at all. Sometimes a heightened emotional experience, even sex, can set one off. In the Boston area, I tell the residents, there's a big spike in these cases in early summer when people start swimming in the ocean. Cold water may be the opposite of sex, but it can create a shocking experience nonetheless, and can trigger TGA. If you have a sense of humor, you get to play the straight man.

"I feel fine. How did I get here?"

The case of the Colombian woman was titillating—a regularly scheduled, amorous encounter that was intense enough to trigger amnesia, but not otherwise noteworthy. It featured the usual stereotyped repetitive questioning with a loss of the ability to form memories going forward. Hannah chose to present it to initiate a discussion of amnesia and its causes, one of the hundreds of neurological syndromes that come up in the course of a year in these morning reports. But to be thorough, I felt our discussion had to go beyond this one case. We needed to consider something less benign, such as the disaster that nearly befell the man from Philadelphia.

At first, Godfrey could remember having set off in the car, yet he remembered nothing of the drive itself. After a few hours, some details came back to him. He vaguely recalled passing through Newark on the New Jersey Turnpike, the smell of the refineries. Under further questioning he could not recall events from a few weeks earlier. He knew the National League standings, but could not remember his last sales call. What bothered me even more was a slight imperfection in his gait. There is no reason in transient global amnesia for someone to have anything but a pure focal memory loss. Any departure from that pattern—the fact that his walk was just a little imbalanced, as though he were tipsy—could point to a potentially deeper problem. I recall him as a very pleasant man, but also as a lovable schlep, neither graceful nor coordinated. He was grateful for small attentions. He savored each one of his hospital meals as though they were the biggest treats he had enjoyed in years. Sitting up in his recliner bed, happy as a clam, he seemed to be a guy who needed caring for. But was awkwardness his baseline, or was it a symptom?

I once treated a woman who was involved in a minor plane crash—a two-seater landed hard and bounced around. Although she did not hit her head, immediately afterward she told the EMTs that she could not remember where she lived or even give her name. For the next few days she turned it into a soap opera. "I roamed around the east coast and I didn't know who I was," she claimed, "and people were so kind to me." In reality, she didn't want to acknowledge an affair she'd been having with the pilot, who was also married. The Blanche DuBois routine was her way of feigning amnesia, but she didn't know how to do it correctly. She didn't know that the one thing people never forget is who they are. She had no idea that this curious thing we call memory works two ways. She would have benefited from reading Lewis Carroll.

In *Through the Looking-Glass*, the second adventure in Wonderland, Carroll tells of a frustrating conversation between Alice and the White Queen. The Queen, it seems, claims to live backwards.

"Living backwards!" Alice says, "I never heard of such a thing!"

The great advantage in it, the Queen replies, is that "one's memory works both ways." When Alice counters that her memory only works one way, the Queen says, "It's a poor sort of memory that only works backwards."

She was right. Memory works both forward and backward. Forward, or *anterograde* memory, is the ability to form memories going forward. Backward, or *retrograde* memory, is the ability to retain memories of the past. The two are inextricably linked: when you lose one, you lose the other.

In Blanche's Hollywood notion of amnesia, post–plane crash, it was possible to lose track of who *she* was while keeping track of who *I* was. She thought she was going to put it over, but if forgetting her own name didn't immediately give the game away (which it did), then "Please help me, Dr. Ropper" sealed the deal. She remembered my name.

I had to close the curtain and say, "Look, I get it. You don't want the story to get out. Why don't we work around that and why don't you stop this?"

She said, "Okay."

Years earlier, when I was a resident at San Francisco General Hospital, I encountered another striking case, this one of true catastrophic memory loss. An ambulance brought in a man in his sixties who had had a massive heart attack, and was unconscious. He had been on the freeway when it happened, on his way to the airport. His companion was a younger woman—the devoted wife?—who followed the ambulance entourage and made it as far as the double doors of the trauma center, then paced around outside, anxiously awaiting the outcome. Inside, the code team descended upon the man, roughing him up pretty good. It was a tough resuscitation.

I was the code leader, and the case required a lot of very fancy dance steps to get him back. When we were done, we discovered that the young woman out in the hall was the man's girlfriend. He had a wife in LA, the girlfriend lived in San Francisco, and he was having an illicit weekend. This would be their last one.

In any prolonged resuscitation, there is a good chance that the patient will emerge with some sort of brain trouble. That was on my mind when he woke up, but he looked great. He was an entertaining, joke-cracking, silver-haired fox of a guy. We told him he'd had a heart attack, and that it was a bad one, and he thanked us for what we'd done for him. "Just happy to be alive." He kept thanking us each time we told him, over and over, what had happened. "Just happy to be alive," he kept saying.

The neurologist who came to check on him was one of my heroes, a compact, perpetually smiling guy named John Coronna, who had been studying the neurological damage done by cardiac arrest and coma. He entered the room with his assistant to administer a standardized research questionnaire that was meant to uncover damage to the medial temporal lobes, the place that seems to serve as the clearinghouse for memories. It is also the area of the brain most susceptible to low blood flow.

John began with the standard questions: "Name? Where are you from? What kind of work do you do?" All went well. The silver fox lived in LA. He was a lawyer. Then John started in on the orientation questions.

"Who is the president?"

"Eisenhower."

Caught off guard, Coronna said, "It's not Eisenhower. *Ford* is the president!"

"What? That's not possible."

"I assure you, it's Ford."

"*Gerald* Ford," the fox said, "*that* idiot? I went to law school with him. He couldn't *possibly* be president."

It was amnesia—Korsakoff's syndrome—a retrograde and antero-grade amnesia, a *permanent* amnesia caused by the low blood flow to his medial temporal lobe during the cardiac arrest. He was now all finished as a lawyer, his memory had stopped, and for him it was 1960. We were stunned. The guy couldn't remember what had happened, could not retain our names for more than thirty seconds, didn't re-member the girlfriend at all, and had no interest in who she was. She was in her thirties, nice-looking, and when she figured out what was going on, she packed up and left.

His wife had to be told, not every detail of course, and I wasn't about to do it myself, so I handed it off to a junior resident, a guy from the Deep South named Lamont Schellerman, who possessed an odd combination of New York cynicism and southern gentility. After a long, heated, and one-sided conversation with the wife, he came back to me and said, "Do I need this?"

From that day forward, the silver fox would live in a world of past memories, unaware that he had a problem forming new ones. By way of compensation, like many Korsakoff's sufferers, he would fill in gaps by confabulating plausible but nonetheless crazy stories. "I think I saw you at the ball park," he might say to someone he had just met. "That hot dog was great, wasn't it?" The urge to fabricate experiences probably grows out of a need to save face. Many alcoholics do it in the early stages of the syndrome, and while it is an interesting component of memory loss, it is not a necessary one.

C. Miller Fisher, one of my professors at Massachusetts General

Hospital at the time Godfrey's car took a spin around Leverett Circle, was the consummate observer. He insisted on reasoning backward from the minutiae of a neurological exam to further his understanding of how the brain works and how disease destroys it. In an obituary I wrote at the time of his death, I called him the grand master of detailed neurological observation. I did not mention the fact that his equanimity was sustained by two extracurricular passions: watching professional football, and the television show *Car 54, Where Are You?* It was Fisher, along with one of my other mentors, Raymond Adams, who had given transient global amnesia its name.

On a Sunday afternoon in November, Dr. Fisher had just settled into his favorite easy chair to watch the New York Giants play the Cleveland Browns when the telephone rang. On the line was an apologetic junior resident who had drawn the short straw, and had been stuck with the job of disturbing the great doctor at home.

"This had better be good," Fisher said.

It seems that the mother of one of the hospital's directors had suffered a fall and was confused. The bigwigs insisted that Dr. Fisher come in and see her. So Fisher came in and did what he did best, observe the patient in a fashion that made him both admired and annoying: he sat for three hours and copied everything the woman said verbatim, and then went on to publish an influential paper about the incident.

The case was notable for the fact that the woman had fallen backward off the chair she had been standing on, had hit her head, and when confronted almost immediately by her daughter-in-law—whom she had known for the past twenty-eight years—had said, "Who are you?" It seemed to be a case of concussive amnesia, but, as often happens, no one witnessed a loss of consciousness. The odd thing was that most of the previous year was gone from her memory, along with selective bits of memory going back over thirty years.

During her examination, the woman was alert and conversed readily, if somewhat hesitantly. She gave correct details of her personal life, including having been born in Lynn, Massachusetts, having left high school in her second year, finishing at night school, then having worked successively at the Preston shoe factory for a year, at the naval shipyard during World War II, at another shoe factory briefly, and then at a Boston bank for fifteen years. She was able to give some details of her marriage, but not of the recent death of her mother and older brother. She gave all of her children's ages minus one year. She identified Kennedy as the president, and when challenged about it, was unaware of the assassination a year earlier. She had not heard of Barry Goldwater, but was able to give a few of the names of her grade school teachers. Most striking, she could not retain Dr. Fisher's name for more than thirty seconds, or even recall having been told it. She kept saying, "I think I've seen you before." She gave the date as six months before the actual date, and many answers were given in an uncertain fashion, usually followed by the question, "Is that right?"

Four hours after their first meeting, she was still unable to retain Fisher's name, but slowly brought into focus the death of her mother and brother. At five hours, she recalled the assassination of the president. One of the most unusual features of the case, similar to transient global amnesia, was that she repeated the same comment with the same inflection each time Dr. Fisher told her his name. "Oh, that's like my maiden name. I won't forget it." (Her maiden name was Fistay.) Other comments she kept repeating were: "What happened to me?" "I think I've seen you before." "Did I fall? I must've fallen on my head because I feel a bump." She improved hour by hour, and after ten hours she was fully oriented to the time, the place, and her situation. Yet on the following morning, the woman could not recall ever having met Dr. Fisher. Her first memory was from ten hours before the concussion, her prior memories having returned, and she could now provide far more accurate details of her early schooling, the principal of her high school, and every one of her schoolteachers.

Amnesia from concussion without loss of consciousness was not unprecedented, and in his paper, Fisher recounted some notable cases in the literature. The most spectacular took place during the Harvard-Yale football game of 1941. When Harvard got the ball to start the game, the quarterback called an incomprehensible play. His stunned teammates naturally failed to execute it, and the team lost yardage. On the next play, the quarterback repeated the same set of signals, with a similar result. On third down, one of his offensive lineman figured it out: The quarterback was calling a favorite play from four years earlier when they were both on the same prep school team. The old play was stuck in his head. "It developed that on the kickoff," Fisher wrote, "the quarterback had received an inobvious blow on the head. By the end of the game his memory had returned, but he remained permanently amnestic for the events of the entire game."

Similar memory losses had been reported by boxers who could remember only a few rounds of fights that had gone the distance. Clearly, it is possible to perform at a high level during an amnestic event, but you might keep calling the same play, or, in a similar vein, get stuck driving round and round a traffic rotary.

By the following morning some of Godfrey's memory had returned, but in a Swiss cheese fashion: there were significant holes in both retrograde and anterograde memory. This is not consistent with transient global amnesia. If Godfrey did not in fact have TGA, he was then, like the silver fox, in serious danger of losing a significant chunk of his past memories, along with his ability to form new ones. The clock was winding down. If I couldn't come up with something, I would have to discharge him at noon.

TGA is highly stereotyped. It varies little from person to person. It is one of the few neurological syndromes that has inviolate borders, and Godfrey's form of memory loss was too spotty, going backward and forward, to fall within those borders. There was also the issue of his awkward gait. I began to worry that low blood flow to his temporal lobes was the true underlying problem, and that he might be at risk of losing a divot out of his brain with a stroke.

In the twenty-third hour of observation, the nurse called me and said that Godfrey's speech had become slurred, and the pieces fell together. I knew instantly that he was having a stroke. When I walked in, his speech was indeed slurred. He was compos mentis, but when I asked to see him walk, I saw that his coordination had completely fallen apart. Godfrey had an occlusion, an atherosclerosis, a garden variety arterial blockage from a cholesterol plaque upon which a clot had formed. As the clot accreted, it had caused decreased blood flow to the temporal lobes, resulting in an evolving stroke instead of a sudden one. It had most likely started to evolve back in Philadelphia around the time he got into his car.

Godfrey's story had a happy ending. We gave him an anticoagulant and an agent to raise his blood pressure, and shipped him up to the ICU. He would be fine, and he left with minimal memory trouble. Had I not held him for observation, the stroke could have cost him much of his long-term memory.

"Be very careful about what you call a TGA," I told the residents that morning. "You're looking for anything that doesn't sound right for a fixed period of *complete* retrograde amnesia and *complete* anterograde amnesia." My guess is that few of them had read Dr. Fisher's paper. That's why I brought it up along with Godfrey's story.

If Godfrey came into the hospital today, the awkwardness of his gait might have been enough to earn him an MRI (which did not exist back in his day), and the stroke might have been evident. Even so, an inexperienced or untutored resident or intern might just say, "No MRI for him. It's just TGA. Let's move him along." Godfrey's was an uncommon condition that mimicked a common one. In the end, it's not really the scan, but the painstaking examination, done Fisher-style, that tells all.

As for the Colombian woman, the residents held her in the Emergency Department for a few hours, but having no memory of why she came, and no awareness that anything was wrong, she insisted on leaving. When Hannah was convinced that her anterograde memory had returned, that it was nothing more than TGA, she discharged her. The hole in her memory would remain, and with it, all memory of her sexual encounter. Fortunately, she had another one scheduled for the following Thursday.

5

What Seems to Be the Problem?

A politically incorrect guide to malingering,
shamming, and hysteria

Her name is Lauren H, age twenty-three, white, brunette, five foot seven inches, 129 pounds. Born in North Carolina, she came to Boston as a student at age nineteen, and is currently employed in public relations.

"I understand that you suddenly became unable to speak this afternoon," I ask her in a rhetorical vein.

"I . . . I . . . I . . . k . . . kkk . . . can . . . can . . . can't . . . t . . . t . . . t . . . talk."

"How come?"

No answer. She stares blankly ahead. Her eyes blink a few times.

Hannah takes me aside and says, "She's aphasic. She must be seizing. Let's put on a hairline." A hairline is a quick and dirty electroencephalogram done with an abbreviated set of sticky electrical leads connected to an EEG machine. The object is to find out if she's in *status epilepticus*, a fancy way of saying that she is seizing uncontrollably. The eye blinking could be a tip-off.

"Okay," I tell her, "go ahead, and I'll keep her talking while you're setting up."

I turn back to the patient. "Is there any reason you may not be able to speak? Has anything unusual or difficult happened to you today?"

A few tears begin to well up in her eyes, and she shifts her gaze away from me toward the window.

"Is it something you can talk about?"

"N . . . N . . . N . . . No."

"Does that mean that nothing unusual has happened or that you don't want to talk about it?"

She pulls the bedsheet up to her nose so that only her teary eyes are showing.

"It's very important that we talk about this because some of the tests that would be done to sort out why you can't speak have risks, and it would be bad for you if we did them for no reason."

A sniffle and a passive look back toward me, but no response. A few absent minutes pass while Hannah gets the material together for the EEG. It's fairly clear that Lauren comprehends me. This would be quite unlike any true aphasia. For one thing, the well-articulated single syllables that stutter up to a full word are very hard for the brain to do. The language areas in her brain must be calling on all of their powers to produce this bizarre speech pattern. From the first sounds out of her mouth, I conclude that it is very unlikely that we are dealing with damage to her brain from a stroke, seizure, or any other acute problem.

Again I ask, "Did anything unusual happen to you today?" Her sister, who has been sitting passively in a chair at the foot of the bed, now pipes up: "She broke up with her boyfriend this morning. Go ahead and tell them, Lauren."

"Was that traumatic for you, Lauren?"

"M . . . m . . . may . . . may . . . b . . . be . . . may . . . be." She sniffles.

"Is that why you can't speak clearly?"

"I don't know."

"That's very good. You *can* speak in a clear sentence. Can you try to speak to me more clearly now?"

By this point a junior resident has finished hooking up all the leads, and the EEG machine is running. It may be a primitive test, but the brain waves look pretty normal.

"Can I get you to say Massachusetts?"

"Mass . . . Mass . . . Mass . . . massmassmass."

"How about Boston?"

"Bos . . . boss . . . boss . . . ton . . . ton . . . ton."

The residents want to get her downstairs immediately for a CT scan and a CT angiogram to see if she's had a stroke. I suggest they may want to slow down and see what happens, but they feel the stroke issue has great time value. They prevail, and she heads off the floor to get a big dose of radiation. She'll be gone for a good hour. I go to see some other patients in the meantime. Later that evening I run into Hannah in the hall.

"What happened with the young woman who couldn't speak?"

"All the studies were normal including her CT angiogram," she tells me.

I manage to resist saying, "Aha!" On rounds the next morning, I ask, "How are you, Lauren?"

"I'm feeling pretty good. Isn't it amazing, my speech came back."

"Yes, it is amazing."

The majority of hysterical symptoms—symptoms that have no basis in disease and are subject to suggestibility—look like real neurological diseases. These include paralysis, inability to walk or speak, blindness, deafness, seizures, and weakness. All are manifestations of an organ that sometimes fabricates problems. But it gets even crazier. People who cannot feel on one side of the body will say they are deaf on that side, or blind on that side, unaware that this is an anatomical impossibility. The hardwiring of the human nervous system cannot produce these defects. This is not disease doing something to the nervous system, but rather the brain doing something to itself. The stomach doesn't have a mind of its own to create stomach problems, nor do the colon, the lungs, or the skin. Ulcers, asthma, psoriasis, eczema, once thought (incorrectly) to be psychosomatic, or originating in the mind, were all shown to have tangible causes, and have been reclassified as nonpsychological diseases.

Only one organ has a mind of its own, and it is constantly causing problems for itself. These problems, once termed "hysterical" and "psychosomatic," are now called "functional" or "somatoform." The conditions themselves are referred to as conversion disorders, implying the conversion of psychic distress into physical symptoms. It is one of the last vestiges of Sigmund Freud's legacy still lurking in mainstream medicine.

Greta B is a thirty-eight-year-old woman, about five foot five inches, well-dressed, a bit overweight. She was referred by a neurologist north of Boston for an unknown type of walking disorder, and has come in with her husband, who is more distraught than she is.

"Hello. Thank you for waiting. I'm Dr. Allan Ropper. I'm at your service."

"Doctor, you are our last resort. My wife just can't function like this. It's been a nightmare."

"Okay. Tell me how it started."

"Oh, it's been going on for a very long time, and it's getting worse."

"How long?"

"What do you think, honey? Maybe six months?"

"Oh, longer than that," she replies. "It's just that it's gotten so much worse, and I'm falling all the time, Doctor. You've got to do something to help me."

"Did it start suddenly?"

"Well sort of, but sort of not," she says, "because at first I caught my toe on the edge of the carpet, and then I began to notice that my legs were buckling going upstairs. That was mainly on the left side, right? And now I can't even walk down the street or get in and out of a car, because as soon as I start, my knees buckle."

"Okay, let me see you get out of the chair without using your hands."

She puts her arms out in front of her, and starts to stand up from the chair, slowly and laboriously, like Frankenstein. Midway through, with her hips and knees bent, she pauses as if starting a dive off a diving board. She then walks forward in duck-step, bent at the hips and knees, like the undersecretary from the Ministry of Silly Walks. When she gets to the doorframe, she grabs both doorjambs, pulls herself to an erect position, and starts to walk down the hall. About every fifth step, her knees suddenly buckle, and she almost falls, like Charlie Chaplin tripping into the sunset, but recovers every time and starts walking again. It puts me in mind of how hard it was as a kid attempting the Cossack dance with my cousins at weddings and bar mitzvahs. Not only does it take a tremendous amount of power and balance to do what she is doing, but it must engage the frontal lobes, cerebellum, basal ganglia, and all levels of her spinal cord. With me in tow, she's doing her own entire neurological examination, and it is super normal.

The effort required to take her by the arm and help her down the hall and then back to the chair, what with the dips and whirls I have to do to keep up with her, wears me out. She plops down into the chair before I can claim it for myself.

"You've just passed the physical, and you're now cleared for the Olympic pentathalon." That's what I want to tell her. Instead I just say, "I think you're going to be okay."

Symptoms are what a patient reports. *Signs* are what a physician sees in an examination. Symptoms are thus subjective, and signs objective. When a patient reports a symptom, we have to take it at face value: a headache, dizziness, numbness, lower back pain. We have no tests for such things, and accept them as real until something in the patient's behavior gives the game away.

The claim of blindness, on the other hand, can be tested. People follow the image of their own eyes in a mirror. Not only that, even if they don't flinch when I bring my hand toward their face quickly, most will involuntarily glance at a $100 bill that I wave in front of them. An old-school neurologist tested this effect. What denomination will get a blind hysteric to follow the bill? A $1 bill doesn't work; a $100 bill works almost all of the time. So he always carried a C-note in his wallet just for that purpose.

———

Lucinda H is a Latina female in her late teens, from Roxbury, with short-cropped and spiky hair, a bit blocky. She is half sitting on the edge of an Emergency Department gurney with her elbows propping her up. Healed slash marks on her wrists stand out against her dark skin. Her mother lurks at the bedside.

"Doctor, she can't see! Oh my God, she's blind, she can't see! *Dios*."

That's the first red flag: Why is her mother speaking for her and why is the girl so calm?

"When did it start?"

"When I came over this morning to pick up the baby. She told me she couldn't see."

"It would be good if I could hear what happened from Lucinda. How about it, Lucinda, what exactly happened?"

"I don't know . . . I'm blind. Isn't that enough?"

"I know it must be frustrating to tell your story over and over to every doctor who comes in, but it's important that I hear the details so that we can get your vision back."

"I can't see anything. How would you feel if you were blind? Tell me that, okay? It's like I'm blind."

"Are you blind in both eyes?"

"That's what being blind means, doesn't it?"

"What were you doing when it started?"

"Nothing."

"Can you see my hand in front of your face?"

"Nope."

"Okay, let me hold your arm and get you up to see how you walk."

I ease her off the gurney, and her feet hit the ground naturally. Without being asked, she makes a left turn and heads for the wall and bumps into it at full stride, but manages to stick her belly out just before hitting, so that neither her head nor knees make any contact. Even before this, I was thinking, *Ho boy!* Now I'm thinking, *Oy, vey!*

I help her back onto the gurney.

"Let's try some other tests. Follow my finger." She stares blankly ahead as I move my finger back and forth. I pull a small mirror out of my bag, and move it from left to right in front of her face about a foot away. Her eyes follow their own images in the mirror. It's gimmicky, like the $100 bill trick, but seeing eyes almost always follow the mirror. Elliott has his own unfortunate variant: he will take a Post-it note, and in small letters write GO FUCK YOURSELF on it, and then stick it on his forehead while he interviews the patient.

"You know," I say to her, "I think you *can* see but for some reason, maybe one you don't want to talk about now, you are just upset or distracted, and are shutting down your vision."

"Well, you're the crazy one, so screw you."

Oy, vey, indeed.

"I'm not saying you are crazy, just that something is bothering your brain in a way that is beyond your control."

Her mother blurts out, "So what are you going to do about it? I can't take her home like this!" Hands are waving all over the room as the mother begins to pace alongside the stretcher, invoking the gods and the prophets at a high pitch. The temperature seems to be rising. I need to leave for a few minutes to cool off.

That's the extent of my plan.

Nomenclature: *Hysteria*, *psychosomatic*, and *pseudoseizure* are OUT. Neurologists still use these words all of the time, just not in front of patients and their families. Other words we take pains to avoid are *psychiatric* and *psychiatrist*. People tend to hear these as *crazy* and *shrink*, and this rarely goes over well with anyone. Terms that are IN include: *conversion disorder* instead of *hysteria*, *functional* instead of *psychosomatic* (the two are not equivalent in any case), and *psychological non-epileptic seizure*—or P-NES (I kid you not)—instead of *pseudoseizure*, as in, "This lady has a P-NES." That's now a term of the art. It was coined either by someone with a very devious sense of humor, or no sense at all.

Susanna B is a nineteen-year-old woman from a devout Pentecostal family living in Plaistow, New Hampshire. She has just started nursing school. She is surrounded by five family members, all seated around her bed, including a hulking brother who has taken the lone easy chair, is fiddling on his laptop, and never once looks up from the screen.

As I rush into the room with the residents, Susanna's arms are shaking, her whole body is shaking. Clearly this is a lovely young woman, but just as obviously, she's in distress. She is fluttering her eyelids at about twice per second, her eyeballs are rolled upward, and her neck is arched backward.

"Susanna!" I say to her. "I'm Dr. Ropper. Can you hear me?"

Her violent movements continue, and her mother eventually stands up and leans over the side of her bed, getting her face as close to mine as she can, and says, "Why aren't you stopping this? We've been here for hours, and we are going to call the patient-care representative. This is unacceptable. We have a lawyer we can talk to."

"You'll have to give me some time to sort this out. It might help me if I could examine her without so many people in the room. Would you mind giving me a few moments alone with her?"

"No way! We're not leaving her alone. Who knows what might happen? She's a virgin, you know."

Wow! That is not high on my list of diagnostic questions, but it raises an entirely new concern. I want them out because this exotic motor performance is characteristic of a pseudoseizure, undoubtedly triggered by the unnerving family dynamic. The quickest way to stop it, I'm convinced, is to send the audience out to the lobby. The neck and back arching, in particular, are not characteristic of most epileptic seizures. Rapid eye blinking can be part of a true seizure, but the way she's fluttering her eyelids suggests otherwise. When the spell finally stops, I speak to her, and find that she is naturally calm, soft-spoken, and quite polite. She knows that the spells are happening, but has no control over them. They were occurring up to several times an hour prior to her admission, and have become totally disabling over the last several days.

From dribs and drabs I pick up from the family, I find out that she represents a great hope for the entire clan. Everything is riding on her success, and they make this expectation very clear at the young woman's bedside. The virginity thing comes up again for no apparent reason, and it gives me the creeps. This is one of the most controlling families I've ever seen.

"Susanna, these are not epileptic seizures. They usually come from . . ."

"What are you saying, Doctor?" The mother again. "We'd like someone else to see her immediately, and if they don't, we will be speaking to people who can make that happen."

"As I was saying, Susanna, it is very important to start by making the correct diagnosis. If we felt that this was an epileptic problem, it would be treated very aggressively with medicines for seizures, but it is not. Somehow, this pattern of motor behavior has gotten into your head, and taken on a life of its own. After a while, movements like this seem to appear without you willing them. Knowing this is very powerful, since it opens up some ways of unlearning the movements, and eventually getting rid of them by retraining the mind."

She tells me that the spells come out of nowhere, have no clear trigger, continue for days, sometimes lasting a minute, other times up to an hour, then suddenly go away. It is very telling, however, that they do not interrupt her sleep, and the video monitor attached to her continuous EEG recording will confirm it. The spells exhaust her, but she lacks any insight into their character. The family gets more and more aggressive, and on two occasions I'm on the verge of offering to transfer her to another hospital, particularly when they suggest they might sue me for malpractice. This is one of the few times that I feel more anger than sympathy toward an anguished family. I take the virgin business, coming up so repeatedly, to be both an admission and a warning: we know that childhood sexual abuse is closely connected to these types of dissociative states, and they are saying, "Don't you dare go there." I'm not going to bother.

Elliott is good enough to come up and try some relaxation exercises with her when the family is not around. He would like to use hypnosis, but there's no way he would be able to get away with it without informed consent. On the third day she is able to whisper and communicate even during the violent shaking and eye-fluttering spells. In between spells, her word choice and mental clarity reveal her to be a very bright and well-read person. Apparently, Elliott tells me, there had been no traumatic episodes in her life, no suicide attempts, no peculiar or borderline behavior, and no family history of psychiatric disease, at least none that anyone will acknowledge. But she's clearly sheltered, if not sequestered in an unnatural way.

The spells stop on the fourth day, and we send her home over the forceful protests of the family that there had been no resolution. They refuse to accept my diagnosis of conversion disorder. I don't hear from them for about five months, when she returns with the same problem, to us no less, requiring another three-day hospitalization.

The Brits call this sort of thing Functional Neurological Symptoms, or FNS, the psychiatrists call it conversion disorder, and almost everyone else just calls it hysteria. There are three generally acknowledged, albeit uncodified, strategies for dealing with it. The Irish strategy is the most emphatic, and is epitomized by Matt O'Keefe, with whom I rounded a few years back on a stint in Ireland.

"What are you going to do?" I asked him about a young woman with pseudoseizures.

"What am I going to *do*?" he said. "I'll tell you what I'm goin' to *do*. I'm going to get her, *and* her family, *and* her husband, *and* the children, and even the feckin' *dog* in a room, and tell 'em that they're wasting my feckin' time. I want 'em all to hear it so that there is enough feckin' shame and guilt there that it'll keep her the feck away from me. It might not cure her, but so what? As long as I get rid of them." This approach has its adherents even on these shores. It is an approach that Elliott aspires to, as he often tells me, but can never quite marshal the umbrage, the nerve, or a sufficiently convincing accent, to pull off.

The English strategy is less caustic, and can best be summarized by a popular slogan of World War II vintage currently enjoying a revival: "Keep Calm and Carry On." It is dry, not overly explanatory, not psychological, and does not blame the patient: "Yes, you have something," it says. "This is what it is [insert technical term here], but we will not be expending our time or a psychiatrist's time on it. You will have to deal with it."

Predictably, the American strategy holds no one accountable, involves a brain-centered euphemistic explanation coupled with some touchy-feely stuff, and ends with a recommendation for a therapeutic program that, very often, the patient will ignore. In its abdication of responsibility, motivated by the fear of a lawsuit, it closely mirrors the beginning of the end of a doomed relationship: "It's not you, it's . . . no wait, it's not me, either. It just is what it is." Not surprisingly, estimates of recurrence of symptoms range from a half to two-thirds of all cases, making this one of the most common conditions that a neurologist will face, again and again.

———

Albert V is a twenty-six-year-old right-handed man, a graduate of a well-known liberal arts college in rural New England. He is well-read, currently works as a chef at a trendy restaurant in Boston. He is thin, pale, freckled, and has reddish-blond, thinning hair.

"Can you make anything of this tremor, Albert?" I ask. Maybe he's not telling me something. What he *has* told me so far is that he had awoken about three months ago to find that all of his limbs were shaking. At first, he thought he might be having a chill, and paid little attention to the problem, even going to work that day. Over the next several days, the tremor got bigger and bigger, to the point where his arms flapped when he held his hands in the air, and his hands slapped the bed when he was lying down. Legs and arms were involved, but no limbs moved synchronously. He had been seen at several hospitals, including one near his home on the Connecticut coast, and he was told he probably had Lyme disease. He is now getting intravenous antibiotics, and is unable to work because of the tremor. It affects his limbs less when he walks, and he finds that he can reach out and touch objects without much tremor.

His speech is normal, and in particular, there's no slurred speech to match the ataxia (his wild and awkward arm movements). His eye movements are also normal. He can walk a straight line, as in a sobriety test, his reflexes are normal, and he seems otherwise fine.

"It's totally baffling to me, Doctor," Albert responds. "It's been getting worse, but there are times during the day when it's much better. My girlfriend tells me I don't have it when I'm sleeping. Is there a chance it's psychological?"

"There's always that chance, but let's see."

With his hands held out in front of him, his arms, elbows, and wrists dart up and down or side to side, arrhythmically but smoothly, like a Hindu devi. When I ask him to put his arms to his sides he looks pretty much like a windmill.

I take his left hand in mine and hold it tightly. The tremor suddenly becomes exaggerated at his elbow. I grab his hand, wrist, and forearm so that his elbow can't move, and now his shoulder goes wild. When I ask the resident to hold his shoulder fixed, Albert's head begins to jerk from side to side. We have essentially chased the tremor up his arm and into his torso. When we let go, the whole arm resumes its crazy up and down, circular and sideways motions.

Next I ask him, using only his right hand, to touch his thumb to his third, fifth, index, then fourth finger, in that order, repeatedly. At first he is unable to do it, but when I take the hand and hold it up in front of him, he starts to make the requested finger motions. As he does so, the tremor in his left hand disappears.

He is an educated and articulate young man. Could this be so opaque to him that he doesn't know what's going on? Could he really think that I don't know what's going on?

A month later, I hear that another neurologist has told Albert that he doesn't have Lyme disease, and the antibiotics are stopped. His parents bring him to a major neurological institute in New York, where they are told that this is a functional problem that requires psychological treatment.

"I'm not saying you're crazy, just that something is bothering your brain in a way that is beyond your control." This is the basis of all non-Irish appeals. "The brain learns these patterns, sometimes they're hard to unlearn, but it's important to know that there's no damage to your brain going on." It doesn't blame the patient, doesn't give a psychodynamic explanation, but does give the patient an out.

"This pattern can be unlearned. It is within your power." That's the moderate tack, not necessarily the American one, because the American sensibility obliges us to add: "Why don't we sit down and find out why you are doing this."

If you pick up any conventional psychiatry book, it will advise you that there is a psychodynamic explanation, or there is a genetic susceptibility, and that by identifying the underlying psycho-problem, talking about it, getting it out in the open, and realizing what it has done to you, you can get rid of the symptoms. If the psychiatrists want to handle it that way, fine by me, but I'd rather not. Besides, most of these patients would rather see a neurologist than a psychiatrist anyway. In their minds, they are sick but not insane. Most of all, they resent the implication that they have a weak character, that they are faking an illness because they can't deal with their lives. I happen to think nothing of the sort, but how do I tell *them*?

Jessica M is a twenty-nine-year-old right-handed woman, who was taken by ambulance from her workplace downtown, unable to move her left side. She is about five foot four inches tall, blonde with dark roots, wears business dress and well-applied makeup. Her right arm is tucked through the handles of her sizable pocketbook.

"I'm so sorry this has happened to you," I say. "Can you tell me how it started?"

"Well, I was coming out of the bathroom and ran into my coworker Nancy. We talked about going out tonight, and I told her I couldn't because I had promised to visit my two aunts who had come to town to see my mother."

"Yes, but how did the problem with the left side begin?"

"Oh! . . . I reached down to pick up my pocketbook, and my arm felt sort of weak, so I asked Nancy to help me sit down on the swivel chair near her desk. Then I felt sort of weak all over, and she told her supervisor to call nine-one-one because she didn't like the way I looked."

"Okay. So your weakness began then?"

"When the EMTs came they told me to grab onto their arm to help me onto the stretcher, but I couldn't grab anything with my left arm."

"What about your leg?"

"When we got here, I noticed that I couldn't stand on it."

The senior resident interrupts: "Dr. Ropper, let's get her going for the CT scan so we can start her TPA." He's referring to tissue plasminogen activator, a powerful clot-busting drug given to stroke victims. Because it is a blood thinner, it poses a serious risk of hemorrhage.

"Give me just another minute," I say, "and I might save us a few hours."

"Show me how you raise your right arm in the air . . . Good. Now your right leg . . . Very good. And you can't move your left leg at all?"

"No."

As she lay on her back, I put my hand under her right heel. "Now push your heel down into the bed really hard . . . terrific."

I move to the other side of the bed, and with my hand now under her left heel, I ask her to try to lift up her right leg while I try to resist. When she does this, I can feel downward pressure in her left heel. In order for anyone to lift one leg up, they have to begin by pushing the other leg down, by way of bracing. This is called Hoover's sign. She is using her left leg without realizing it.

I explain this physiological fact to Jessica, but before I can finish, she exclaims that she feels her left side getting stronger.

I tell the residents not to bother with a CT scan.

In many cases of hysteria, the ideal treatment would be hypnosis. We used it when I was a resident, and it worked, just as it worked for Sigmund Freud and his teacher, the French neurologist Jean-Martin Charcot over a century ago. It worked because patients with hysterical symptoms are suggestible, and, having fooled themselves *into* the symptoms, they can be fooled out of them. Deception works, but in the modern age, in the age of informed consent, we are not allowed to fool patients about anything, even if it is the only way we can help them.

———

Victor P is a twenty-seven-year-old Russian émigré, an on-and-off student, and peripatetic barfly. He has been seen in the Emergency Department by the neuro-consult team, and is still there, awaiting a diagnosis.

At 7:30 a.m., I call the residents to order at morning report. "What have we got?"

"Victor," Hannah says. "He's been in the emergency room since about five a.m., when the EMTs brought him from his apartment down on Huntington Ave. His roommates said he was having weird movements and foaming at the mouth."

One of the junior residents, obviously thinking back to his college fraternity days, adds: "Can you believe it? This started at about three a.m., but these jerks thought it would be fun to watch him do this for a while. I think they're BU graduate students."

"Just now, when I left the ED," Hannah adds, "he was shaking all over with wild, rhythmic movements that lasted for about forty-five minutes. We've already tried Valium IV two mg, Dilantin IV up to a gram, and they're breaking out the Versed."

"Hold it," I say, "that's pretty potent stuff. What makes you think that forty-five continuous minutes of shaking and shimmying has to be a seizure? Did he bite his tongue?"

"No."

"Was he incontinent?"

"No."

"Look, seizures stop themselves after a couple of minutes. The cells exhaust themselves and use up all the ATP, so it's almost impossible to convulse for this long without stopping and starting again. Is his back arched?"

"Yes."

Elliott, seated to my right at the conference table, suddenly stands up and walks out of the room without a word. He returns almost immediately with a rubber stamp in his right hand, reaches over my shoulder, plunks an inkpad onto the table, inks the stamp, smacks it down next to the case number in the log book, and sits down in a huff.

It's a little red turkey.

Pseudoseizures, or P-NES, are probably the most common form of hysteria or conversion symptom seen in a neurology ward. The majority of them occur in patients who have genuine epilepsy. They know what a real seizure should look like, and are pretty good at producing fake ones. The siblings of patients with epilepsy, who have seen plenty of real seizures, also account for a significant number of cases.

Victor had no previous epileptic history, but we have to admit him to the ICU and take continuous EEG recordings for a day or two just to confirm that nothing shows up on the recordings. Nothing does. Moreover, he "wakes up" from his long convulsive sessions perfectly lucid, a virtual impossibility after a true epileptic spell. The facts are duly noted in his record for the benefit of the next hospital that has to deal with Victor.

The average American has at least one unexplained symptom every week or two, and less than one-fifth of survey subjects report no symptoms at all during the three days prior to a random query. Headache, tingling, pain, dizziness, briefly blurred vision, a slight imbalance when walking, loss of train of thought, feelings of jabs and jolts: these are the most common symptoms, and they befall any healthy nervous system, then disappear and are forgotten. In some instances, however, there is no way to disabuse a patient of his or her symptom. Where a normal, functioning person might say, "Well, things happen," some patients develop an attachment to or an excessive worry about a specific extrinsic cause of their suffering. They fixate on it.

Depending on the decade, it might be electrical fields or Lyme disease, environmental allergies or chronic candida infection, hypoglycemia or sick buildings, perfumes, or even alien abduction.

The trick is to communicate the fact that you know, and that the patient needs to know that you know, and the family also needs to know, that the problem is not organic, somatic, neurological, medical, life-threatening, contagious, critical, or cataclysmic. That is not to say that it isn't disabling in its own way. Hysterias and factitious disorders prevent people from getting on with their lives, and prevent their loved ones from getting on with theirs. They are highly disruptive. Whatever strategy we choose to take is obliged to acknowledge that it is an illness like any other, and work from there. But there are limits to what I can tolerate.

Elliott slides a letter across to me, and then resumes his mandarin pose: head bent forward, eyes closed, elbows on the padded arms of his desk chair, hands interlocked in the church-and-steeple grip, index fingers slightly indenting his pursed lips.

It is a long letter, and Elliott holds his pose while I scan it.

"Well?" I say.

"I refuse to see them," he replies absently. "Might as well open the floodgates."

"But this is something rich, don't you think? Aren't you curious?"

"No. It's loony tunes."

"Well, I'll see them," I say. "In fact, I have just *got* to see this."

On the appointed day, I take the couple to one of the examining rooms. Prescott K, age thirty-five, who wrote the letter, and his wife, Lizette, age thirty-three, have come a great distance with an electrifying story. They are clearly prosperous, very well-dressed. He's in the hedge fund business, and wears khakis and loafers without socks. She is thin, not unattractive, and is wearing a stunning Chinese jacket of red silk, with a traditional collar, black piping, and a magnificent

Hermes scarf. Prescott immediately looks around and backs himself into a corner of the room.

"What are you doing?" I ask him.

"Well, the outlet there, and your computer screen are making me extremely uncomfortable."

"How do you mean?"

"It's that buzzing feeling. It's what I came to you for. There's an internal vibration that's extremely violent, and goes up to my head. Let me explain, Doctor. I got this problem first, even before we were married three years ago. I began to notice that my whole body buzzed when I walked by electrical outlets or appliances when they were on. After living in our new house for six months, my symptoms got much worse, and my wife got exactly the same symptoms. I had electricians rewire the house, but it's still happening. I know what you are going to say, but we are not crazy. I didn't have this in my first house with my first wife."

I know something about electrical fields, I know something about neurology, and I know something about baloney. Baloney is not hysteria or a conversion reaction. Unlike conversion, which has everything to do with overt neurological dysfunction, the more outrageous the better, baloney involves the exaggerated reporting of personal neurological experiences that is focused on nonsense, usually abetted by the hyperbolic popular press, all bordering on the delusional.

Schizophrenics, who suffer from something tragically real, have manias that have kept pace with technological advances. First, voices spoke to them from the dark, then voices came out of the radio, then voices came out of the television, now voices come out of the computer, from the Web. This much they have in common with the electrical couple: the aggravating factor is the world itself, if not the universe. They can't tolerate the world we live in. But while the schizophrenic can't pick his demons, for the electric couple it's a lifestyle choice.

Of course, sharing the illness du jour with large numbers of fellow-sufferers has a supportive, salutary effect. In this case, the affliction is called electromagnetic hypersensitivity, or just electrical sensitivity syndrome. There is a wealth of Web sites and chat rooms devoted to it. Sadly, when it comes to dealing with, much less treating, such borderline theories, I have no spiel to offer, and sometimes revert to being a jerk. In this case, I suggested that they both might be magnetized. As an experiment, I said, he and his wife should float on their backs in their swimming pool to see if they both pointed north. I was guessing that they had a pool. I was right.

They never came back.

The malingerer is not an hysteric. He (it's usually a man) is purposely putting on a show to dodge the cops, avoid jail time, get out of work, receive compensation, or score some drugs. Likely he's had practice, and has been to a couple of emergency rooms where he has honed his act like a stand-up comic.

Ethan R is a twenty-six-year-old graduate of Brandeis University, where he played soccer and majored in labor relations. He is unemployed, overwrought, increasingly desperate, and possibly in pain. It is hard to tell.

The residents are appalled. Without saying anything, it's clear what they are all thinking: "Why did you admit this guy, Dr. Ropper? He's nothing but a drug seeker!" Thus have I revealed, on only the second day as the attending, a fundamental weakness: my first instinct is to take what people tell me at face value.

In the outpatient clinic, Ethan told me that ever since falling off a ladder a day earlier, he had been suffering such excruciating sciatic pain that he couldn't sleep, couldn't even find a comfortable resting position. Now perched on a bed up on the ward, in a room he shares with a heroin addict, Ethan becomes a method actor. As I examine him, his eyebrows circumflex into an emoticon of anguish. The whole shtick is way over the top, and I see some of the residents roll their eyes. What he wants is precisely what the pain service refuses to give him: intravenous Dilaudid and a prescription for oxycodone and Percocet.

Two things give him away as a drug seeker. The first is his preemptive insistence that he is not one. The second is his familiarity with drug names and dosages. His frustration seems to stem not so much from the pain as from our refusal to give in to his requests for drugs.

"We're not going to do very much," I tell him. "We'll see how you're doing tomorrow, but let's talk about the drugs up front, now, because I saw you in the office, and I thought you were a pretty straight shooter. I don't want there to be conflict about it. What we need to do is to start to taper the drugs off, and move things toward something that's a little bit more manageable as an outpatient."

"I understand what you're saying. I'm not going to ask you for any medications to take home. The thing is, the way it starts and it stops . . . there's no decline in pain, there just *isn't*."

"So you mean that once you get into one of these cycles of pain, it's just there?"

"Yeah. I mean at first, it's just . . . it's hell, it's hell on earth."

"But you can still walk out to the elevator and go out and have a smoke?"

"Yeah, I can. I could at the beginning. With help."

"But I've seen you walking around. I know you're uncomfortable, but . . ."

"I have a very high tolerance for pain."

Such a high tolerance that, at the end of the workday, Elliott sees Ethan in the lobby, perched on the edge of a sofa, laughing it up with one of his buddies as he casually munches on a fruit cup.

The following day, the family comes in—both parents—very concerned. In Ethan's room, Elena, one of the junior residents, gives them "the talk":

"Mr. and Mrs. R, this is a problem that needs to be addressed before moving on, but it can't be properly addressed or solved in the hospital because we don't have the tools here. You need psychiatry, you need physical therapy, and in fact, we're doing you a disservice by having you here in the hospital at all. The best thing we can do for you is to hook you up with these amazing services that could actually help you."

Brilliant! I'm thinking, even though the message has a depressive effect on Ethan, not that he even matters at this point. What matters is his parents, because they are inadvertently enabling and facilitating this.

Elena makes the phone calls to set up the appointments, we offer salves, we offer some therapy, but in the end, just to get rid of him, Elena has to bribe him with a shot of intravenous Dilaudid. That's what he came for, and that's what he gets on his way out the door.

Children under about the age of eight do not get hysteria, probably because they are guileless. On the other hand, if a young woman, especially an adolescent, is suspected of hysteria, a pink bunny or teddy bear next to her pillow is a tell. It is almost a guaranteed sign of a conversion disorder.

First I knock, then I enter. I see that the shades are drawn, the lights are off. It is not unusual to walk into a patient's darkened room. If somebody has meningitis or a subarachnoid hemorrhage, the light truly hurts their eyes, and the nurse will have posted a cautionary sign outside the door. There's no sign on this door, leading me to believe there will be a messy psychological situation waiting for me inside.

As my eyes adjust, I see a young woman on the bed, wearing pajamas with pink slippers, hugging a large pink teddy bear with a bow in its hair. I also see her mother, eyeing me suspiciously from the lounge chair at the side of the bed. My snap diagnosis: conversion disorder, pain, dysfunctional family situation, possible childhood sexual abuse.

"Hi, I'm sorry to intrude. I'm Dr. Ropper, and I'm in charge of the neurology service, so I will be your doctor."

The mother stands up immediately and inserts herself between me and the foot of the bed. She is holding a stack of papers in manila folders under her left arm, and she takes her glasses off her nose, letting them dangle on the chain in front of her.

"She really doesn't want to see any doctors right now. And she hasn't gotten her Demerol."

"Maybe I can examine her before we get some Demerol."

"I don't know why you simply can't read her records. She has reflex sympathetic dystrophy, and she needs her pain medication now."

I try to poke my head around her mother's shoulder to peer at the patient. "That's a nice teddy bear you have."

"I take him everywhere."

The mother shuffles to the side of the bed, level with her daughter's waist, and will not let me go any further.

"We go through this all the time. New doctors thinking they have a better solution and trying to get her off her pain medicine."

"We go through this all of the time, too," I'm tempted to say. "Mothers who poison their daughters so that they can nurse them through recovery."

What I wouldn't give to be Irish right about now.

6

Do No Harm

A walking time bomb tests the limits
of good sense

It was a hazy, humid, and windless mid-August afternoon—wicked hot, as the locals like to say. At a car-strewn front-end shop on Huntington Avenue, Ruby Antoine rolled under a Monte Carlo and set to work on the transmission. Because the garage had only one working lift, and it was occupied, Ruby had the front of the car jacked. As he was pulling the drive shaft out, at the moment of maximum strain, he felt the sudden onrush of a headache so intense that he thought his head would explode. Barely able to roll himself out on the dolly, he summoned all of his strength to haul himself up against the standing tool cabinet, where nausea overtook him and he vomited clear across the work bay.

"Holy shit, Ruby," his boss said. "What's going on?"

"Wicked headache, man. I have no idea where that came from."

"We've got to get you to a hospital."

"No, man. I'll be alright." But he sure as hell didn't look alright.

By the time his objections had been overcome and he had arrived at the emergency room at Boston City Hospital, the headache was almost gone, and a casual exam turned up nothing. Ruby was so evasive and in such a hurry to get home that the intern became suspicious, thinking it was probably drugs. The doctor there ordered a CT scan as a routine precaution. It came back normal, and Ruby was discharged. He was fine, or so he thought.

Four days later, Ruby was at home on the couch playing with his kids and watching TV when the same thing happened. He again refused to go to the emergency room, but as he sat clutching his head, his wife grabbed the keys to the car, took his arm, and led him outside. He was so incapacitated by pain that for once his wife prevailed, and this time took him to the Brigham, which is only blocks from their house.

It is a telling fact, entirely consistent with the sad history of race relations in Boston, that despite the predominantly African American composition of its immediate neighborhood, there are not too many black faces in the crowd at the Brigham, a trait it shares with Fenway Park just down the road. But Ruby's wife had no use for such folkways, and didn't want Ruby getting sent right back out the door. So she brought him to us.

After hearing the history, the Emergency Department resident got yet another CT scan while waiting for a neurological consult. To the resident's relief, nothing showed up on the scan, but I was less sanguine. "I'm really worried you've had a subarachnoid hemorrhage," I said to Ruby at our first meeting. "Even though it doesn't show up on the scan, we need to do a spinal tap."

A subarachnoid hemorrhage, or burst aneurysm, is one of the most treacherous diseases in medicine. Although considered a type of stroke (or to use the Boston vernacular, a "shock") because it happens suddenly and occurs in the head, it bears no resemblance to the paralysis or speech trouble usually associated with a stroke. An aneurysm is a rounded pouch on one of the blood vessels at the bottom of the brain. If it reaches a critical size and form, it can burst open with the entire force of the body's blood pressure. Blood then fills the spaces around the brain in a split second and causes a thunderbolt of a headache that no one forgets and many don't survive. One-third of patients who do survive have a re-rupture in days, and half of those people die. Ruby had dodged the thunderbolt twice. He had had two bleeds, which resolved, although his blood had now mixed with the spinal fluid surrounding his brain—not enough, apparently, to show up on the CT scan (in about 5 percent of patients it does not), but enough for us to confirm it if we could get a lumbar puncture.

"No way," said Ruby. "That's not gonna happen. No way somebody is sticking a needle in my spine." And that was just the beginning of a long and frustrating story.

There are three kinds of patients who show up on the ward: the risk-neutral, the risk-averse, and the risk-resistant. We rarely see a fourth kind, the risk-takers, who instead show up in the morgue, if they show up at all. Most of the time they go straight to the funeral home. Risk-takers don't come to the hospital of their own volition. They simply refuse to go to a doctor for anything. By contrast, risk-neutral people, a term that describes most of our patients, go to the doctor because they believe in modern medicine. They listen, they question, they comprehend, and for the most part, they cooperate. Risk-averse patients, on the other hand, need reassurance. These are the people who have to know everything. They have to call everyone. They need a better opinion. They won't have the operation until they've had five second opinions. They anguish, they're neurotic, but in the end they do what all normal people do when confronted with choices. Then there are the risk-resistant, who are simply recalcitrant, and give us the most trouble. They act as if they have something to hide, they often do have something to hide, and they hide it even when it would be in their best interest to let us know about it up front. Ruby Antoine fell into this last category.

One way to deal with the risk-resistant is to have them sign a piece of paper acknowledging that they understand the risks of doing nothing, then send them home. The problem with Ruby was that, as enterprising and personable as he seemed, he had never learned to read or write. Ruby Antoine was a rangy, muscular, engaging, yet infuriating man. He had survived and even thrived in the literate world through a combination of obstinacy and charm. Instead of withdrawing into himself out of embarrassment or shame, he overcompensated by developing hands-on skills—he fixed cars, built stone walls, and had even learned blacksmithing. A jack-of-all-trades, he had grown up on Cape Cod, had had very little formal schooling, and had navigated his way through life using his ability to recognize food labels and road signs as symbols rather than as text. He was unimpressed by me and skeptical of what I was telling him. Not only was he not going to agree to the tests I wanted, Ruby was not going to sign a waiver. He wasn't even going to glance at it.

Those who know anything about the oath of Hippocrates, which is still recited at some medical school graduation ceremonies, usually recall its main precept: *primum non nocere*—"First, do no harm." For many physicians, this is not only a guiding ethos, but a justification for conservatism or, more precisely, for doing nothing beyond prescribing aspirin and bed rest. But the Hippocratic dictum in the modern era creates an unsettling conundrum: doing nothing with the knowledge we have in a case like Ruby's risks tremendous harm. The risks of failing to make the diagnosis of a ruptured aneurysm are far greater than the risks of a spinal tap, an angiogram, and doing an operation.

If Ruby had been a "good" patient, a risk-neutral one, he would have agreed to the spinal tap after a bit of reassurance that my residents have done hundreds of them, even on little old ladies, and while he might have ended up with a bad headache for his troubles, it would have given us some vital information. If the spinal fluid revealed the red-brown color typical of broken-down blood cells, it would have confirmed the diagnosis. After that, an angiogram—essentially an X-ray that reveals the outline of the cerebral arteries—would have

isolated the aneurysm, and led to a rapid treatment. Instead Ruby
squandered time, resources, and a considerable amount of goodwill
by sitting there and refusing all of our attempts to either move his
case forward or move it out the door.

"I just want to go home," he said.

"But you're a walking time bomb."

"Yeah, and I'm walking right out of here."

He certainly had that right, but before throwing in the towel, I
reached out to Edgar St. Claire, one of the hospital clergy, whose somewhat easy manner belies the fact he takes himself and what he does
very seriously. Hospital ministering is a special corner of ecumenical
life. End of life, fearing the end of life, having a loved one near the end
of life: this is the proverbial foxhole that lacks atheists. In our foxhole,
there are on staff three priests, a female reform rabbi, a Baptist minister
(Edgar), and several denominations on call, including Greek Orthodox,
Muslim, and Orthodox Jewish. I have not seen Baha'i but I'm pretty
sure we could find one.

Hospital clergy are schooled in the human spirit, whereas I'm
schooled mainly in the human body. They are more sensitive than I to
a patient's belief system, cultural context, and motivations. Edgar in
particular is the ultimate humanist. When he comes into a situation,
there is no right and no wrong. "I'm here to administer to the soul
and spirit of this individual," he has told me, "so I have to understand
who this person is." I had to concede that I didn't know who Ruby
Antoine was, and I needed help with that.

"Edgar," I said, "I have this problem. I think this guy is a danger to
himself. I don't want you to talk him into the spinal tap. I just want
you to get him to reconsider the whole situation." So Edgar and I went
to see Ruby to hear why he didn't want the tap.

"Because I don't want to get paralyzed . . . because I got better the
first time . . . because I'll get better again. I ain't worried I might die of
some burst aneurysm I may not even have."

Ruby was never angry or sarcastic, merely insistent. He was difficult, but in a good-natured way. He let you know there was nothing that he couldn't do, and he was ready to prove it. Yet intentionally or not, he was pushing all of my buttons, and I might have blown a gasket had it not been for the sheer absurdity of the situation, or at least one very small and entirely irrelevant detail: I couldn't get over his hairy legs. Most men lose their hair below the sock line, but Ruby retained thick, bushy hair all the way down to the ankles, even on the back of his calves. Whenever I visited, he tended to lean back on his left elbow and drape his legs over the side of the hospital bed, a languid posture that made him speak out of the right side of his mouth and upward toward the ceiling like a Bedouin warlord. All that was missing was a hookah and an oriental carpet. He was anything but effeminate, but when he crossed his right leg over his left *below* the knee it left a peculiar impression. It isn't natural for most men to cross their legs that way, and it made me wish they had found him a gown of the right size, or, even better, a pair of blue hospital pants.

I introduced Edgar, and left the two of them to it. Edgar spent an hour with Ruby, came out, and said, "I did what I could, but in the end, you have to respect his wishes as a human being." I went back in and gave Ruby a long look, and thought, but did not say, *You are a schmuck.* I would bet that Ruby looked at me and thought more or less the same thing.

Edgar came back the next day and the day after that, and the two men went through the same dance with the same result. "We're trying to save your life. You've got a wife and two lovely children. They need you."

"Well, they got me. And I got to get back to work 'cause I ain't making money sitting here."

On the third day, after he was discharged, I was still having trouble accepting his wishes as a human being.

Edgar St. Claire is someone you notice—short and square, but not stocky; broad shouldered, but just as broad at the waist. At five feet five inches, with a plodding and wide-based gait, and almost always carrying a briefcase or folio under his right arm, he is easily spotted from two hundred yards away. He most often wears dark brown or black suits of a noticeably elegant design, with well-chosen ties of colorful fractal patterns, often liturgical purple. With his head held erect or tilted slightly backward, he exudes a stately demeanor while avoiding intimacy. His most arresting aspect is the timbre and pattern of his speech. His accent is clearly Caribbean, but measured in its excess so as to allow a British lilt to emerge from the phonation—with a well-taken breath halfway through each sentence, just for emphasis.

"Edgar, how did you get into this business?"

It was the following day. I was trying to process what had happened, or not happened. I wanted a story to distract me, and Edgar gave me one.

"You know, in Jamaica, when I was young, I was not too much involved with the church. I had too many folks in my family that were Baptist ministers, and I didn't like the idea of having to depend on people. I always wanted to be able to work and earn and do what I wanted to do. My goal in life was to be a private detective, or at least to be a good businessman. But my brother and I didn't know the first thing about how to be detectives, so we opened a convenience store. Just to go in and hear the cash register opening, even if it wasn't to put in money, to me that was good. In 1972, I attended a church college, and that spoke to my heart about ministering. I resisted that for a while, but the Bible says the calling of God is without repentance. If God calls, you are never satisfied until you find yourself doing what you feel you are called to do. That's where I found myself. So I went to church bible school, then I went to Mississippi Baptist Seminary, and in our church they move you around, so I stayed at Mississippi for about twelve years, then Tennessee and Georgia for a while. I've been up here for about twenty-seven years."

"Did you tell Ruby all that?"

"Yes. We talked for a long time. He asked me about the hospital, about what I thought of it."

"What *do* you think of it?"

"Well you know, I was a patient here. I had surgery here ten years ago. Not only was I a patient here, but I thought so much of this institution that I pursued a job here."

"But as a black man . . ."

"Well, there were problems. When I came here for surgery, I was on the sixteenth floor in the Cardiovascular Center, and no one from the clergy came and visited me for days. I remember saying to my wife, 'As an urban institution sitting in the middle of where we dwell, how can they not have anyone that looks like us as a chaplain here?' So I decided that I'm going to have to put some things in place to see if I could change that."

"And since then?"

"I feel that it has changed tremendously. I think going out into the community and having clinics and hiring minority doctors and nurses has helped. I'm finding that there are more African American patients coming than when I first started."

Still wondering what had gone on with Ruby, yet still reluctant to ask, I kept up an oblique line of questioning. "Do people ask you about deep existential, theological issues or problems? Do they ask you: Is there a God?"

"Many times. You always have that question: 'Is there a God? And if there is a God, why am I going through this?' Back when I was the chaplain on duty, just starting part-time, a young lady was med-flighted here. She was pregnant. They had to rush her into surgery, open her up, take out the baby, all wrapped up in a blanket, and they gave it to me to bless because they knew the baby was not going to live. And one of the nurses said, 'You need to go in there and pray for the mother.' She was coding, blood was going everywhere, and this was the first time I'd ever seen someone with a human heart in their hand, pumping it to try to keep her alive. I blessed the baby, I prayed for the mother. She lived, but the baby died. That stuck with me. I met another young lady just the other day. She had lost her daughter a few months earlier. Her husband was in the hospital, and his prognosis was not too good. She said, 'I'm mad as hell with God. As a matter of fact, I don't want to hear nothing about God.' And this is a person who professed to go to church. So there are times when you have to be able to let them pour out exactly how they feel without condemning them. You don't have to prove to them that there is a God. Sometimes it's just what it is. They have these questions—'I'm serving God. Why am I sick? I've been a good person, why is this happening? Why has my child died?'—I don't have the answer to a lot of these things. I sit there and let them pour it out, and sometimes they say, 'Thank you just for listening.'"

"And Ruby . . ."

". . . he has his reasons. Let's leave it at that."

As it turned out, Ruby did have some good reasons, but I would not discover them until it was all over. And we still had a long way to go.

A month later, Ruby's primary care physician talked him into getting a spinal tap at the Brigham. I have no idea how he managed that, but by the time it happened, whatever blood had been in the spinal fluid would have been mostly reabsorbed. I did it anyway, and the spinal fluid did show the remnants of old blood, indicating that he had indeed had a subarachnoid hemorrhage. I then miraculously succeeded in persuading Ruby to have an angiogram, and the angiogram revealed a very large, multilobed, anterior cerebral arterial aneurysm that justifies its name: "berry aneurysm." Moreover, in Ruby's case it was like a mutant blackberry rather than the usual small blueberry. There it was in black and white, a bulging blob hanging off the artery, ready to explode.

"I don't want it fixed."

Here we go again, I thought. I explained to him that now he had actually left the starting line. If he stopped in the middle of the race and decided that he wasn't turning back, then he'd just be stuck in the middle of the track with something big bearing down on him, and then he'd really be asking for it. Fed up with niceties, I said as much: "Then you're screwed, Ruby!" But he wouldn't listen, and again called his wife to pick him up from the hospital.

"What's with this guy, Edgar?" I asked the next day.

"Well, he has his reasons," he replied, somewhat obtusely. "We just have to accept them."

Three months later I learned that Ruby's primary care doctor, who must have had Periclean powers of persuasion, succeeded in convincing him to have the surgery—finally! Unbeknownst to me, he ended up at another hospital, where the operation almost wrecked him. It happens, even with the best surgeons.

The definitive cure for an anterior cerebral aneurysm is an extremely delicate procedure. It involves the placing of a clip that looks like a miniature distorted clothespin on the neck of the aneurysm. The trick of it lies in choosing the right clip, choosing the right place to put it, getting just the right angle, and making sure that you seal off the neck of the aneurysm so that no blood can get into it, all the while avoiding all of the surrounding vessels. *All* of them. The clips are made with little cutouts, allowing them to put pressure only on the aneurysm and not on any surrounding blood vessels. But in that area of the brain it is very easy to nab tiny vessels that feed the frontal lobes. When the surgeon put the clip in, he probably nipped a few of these and turned Ruby Antoine into a Randle McMurphy: a dull, frontal-lobotomized kind of guy, a guy who can never go back to work. He was in his late thirties at the time, with two kids.

A few months later I ran into a colleague who had at one time treated Ruby. "Did you see this guy?" he asked.

"Oh, boy, did I see this guy. I spent hours trying to convince him to do something."

"So did I," he said. "I guess the patient's always right."

And yet the lesson isn't that he was right. Looking at Ruby's angiogram even now, I can state with a high degree of certainty that he would have died had he bled again, and he most certainly would have bled again with that kind of aneurysm. I would still give the same advice to anyone in Ruby's situation. It wasn't a question of choosing the wrong hospital. He was operated on by an extremely capable surgeon. There is an alternative to clipping called endovascular coiling, in which the surgeon gets at the aneurysm via a micro-catheter fed through the artery, but that procedure has problems, too.

Six months later, I screwed up the courage to ask Edgar directly why Ruby had refused all counsel when he first came here, and what changed his mind.

"Oh! Well, there was some question in his mind whether you were going to use him for an experiment, as a guinea pig." This hit me like a ton of bricks. Later that afternoon, as Elliott and I sat in my office reviewing the events of the day, I told him what Edgar had said about Ruby. I still couldn't wrap my mind around it.

"Haven't you ever heard of The Plan?" Elliott asked.

"Vaguely."

"It helps if you lived in DC during the Marion Barry years. I was at Georgetown back then. Everybody knew about The Plan. Supposedly, it was an effort by whites to retake control of city government. Maybe some whites did have a plan like that, but it was mostly a figment of the black press, just one piece of a broader conspiracy theory. Black genocide. You know: the World Health Organization created the HIV virus in the 1970s, the CIA used it in an experiment in Africa while they pumped drugs into the black ghettos here at home. That sort of thing. Abortion as an anti-black eugenics plan promoted by Planned Parenthood. Then you have the real stuff like the Tuskegee syphilis experiment."

So Ruby wasn't too far off base. The Tuskegee syphilis experiment was one of the most notorious clinical studies ever conducted. From 1932 to 1972, the U.S. Public Health Service offered free health care and meals to rural black men in return for being allowed to follow the progress of their untreated syphilis, without informing them of their disease or the existence of treatments to cure it.

"Of course," Elliott added, "we do experiments on people here all the time."

"With informed consent."

"Yes, but that's a finer point that might understandably be lost on your patient. Seems to me his fear was a pretty rational one."

"Well," I conceded, "I never would have guessed it in a million years."

"What we've got here," said Elliott, quoting *Cool Hand Luke*, "is failure to communicate. You had three parallel belief systems going on: yours, the rational, scientific belief system, the one that can't even conceive of something like The Plan; Edgar's faith-based belief system of a learned man, but not a man of science; then you had Ruby's. Unless you could spend some time in that front-end shop on Huntington Ave, maybe hang out with his friends, put yourself in his shoes, I don't think you're going to get much insight into his belief system, but it was as legitimate to him as yours is to you."

It occurred to me that Edgar is willing to hold two competing ideas simultaneously: science is a good, medicine is a good, but there's also something beyond it. Then, as Elliott pointed out, there's Ruby and his belief system. He's not comfortable even stepping into this hospital. His comfort level going to Boston City Hospital is much higher because he knows, or at least believes, that they're not going to experiment on him.

"You may have put some doubt into that worldview," Elliott said. "Edgar told me that Ruby liked you, put some trust in you. Maybe that's why he finally caved and agreed to the operation. Where is he now?"

"I don't know," I said. "All I know is that he's alive, which he wouldn't be if they hadn't operated on him."

"So did we do no harm?"

I had no answer. Medicine is like the laughing and crying masks of the theater—comedy and drama. There is never one side that is right and another side that is wrong. The risk calculations are in the doctor's head, and no algorithm has yet to do better. Yet ultimately, as far as the hospital is concerned, the patient is always right because personal autonomy trumps probabilistic outcomes. You have to respect their wishes as human beings, we are told. But if you ask me whether the customer is always right, I would say, "Not at all." The patient is so very often dead wrong, and very much so when it comes to his own brain.

7

A Story Is Worth a Thousand Pictures

Nine songs of innocence and experience

1. LOOK, LISTEN, FEEL!

On the first day of my second year of medical school at Cornell, I was sitting in the next to the last row of a small, steeply sloped auditorium. A legendary cardiologist and internist named Elliot Hochstein was on stage preparing to teach a course on physical diagnosis: how to properly examine a patient. For some reason, the lighting in the back of the hall didn't work quite right, so the stage, the speaker, and the first five rows were always in the light, while the back rows, though not dark, were dim. It reminded me of a baseball stadium when the sun casts a shadow across the infield. I used to say, "They need to put a foul pole in here."

There were eighty-nine of us in the entering class, and we were all in attendance. It was mandatory. Next to me sat a woman named Judy Vanak, and next to her a hippie-type guy named Roger Stuxman. Almost four decades later I can recall Judy very well because there were only five women in our class. We had more Mormons than women. We had more hippies than women.

Ten minutes into the lecture I was startled by the weight of a heavy hand on my shoulder. It belonged to a guy with very long hair, who climbed right over me, sinuously propelling himself like a cat into the next row, then over that row, and onward to the front. As he made his way forward, literally walking on all fours over one shoulder after another, our attention was drawn to a second, third, and fourth wave of the invasion. Several women crawled over the backs of our seats, followed by more long-haired men. The weirdest thing I had ever experienced. Alarm gave way to annoyance, then astonishment. They swept over the entire entering class, crossing from the dark into the light, more than a dozen of them, and assembled loosely at the front of the auditorium, where they started belting out the anthem, *Hair*.

It was the Broadway cast.

Hochstein came up with the idea. The show's producer was his patient. They did the entire musical number, a mind-blower, and after they left, Hochstein said to us, "What do you think?"

A guy named Tory, a very talkative kid who always sat in the front row in order to ask inane questions (and who, of course, went on to become a Beverly Hills gynecologist), said, "Geez, that was an incredible experience."

And Hochstein replied, "Exactly! The trick to medicine is to have it be an active experience, not to be a passive observer. Get in there fully with your senses, and then you'll be a great clinician. Look, listen, feel! Don't just stand there. Leave a big mental opening, because if you go into an encounter knowing what you're going to find, you *will* find it, and yet you'll miss the important stuff. And the important findings when you are examining a patient are multimodal. It's not just: you hit the knee with a reflex hammer and the leg goes up this high, maybe too much, maybe too little. No! It's the whole ensemble. It's all the senses engaged together."

He was onto something, because the actors had a very physical presence. He was talking animatedly about the medical examination, yet he was also talking about the performance we had just seen. When they went from the dark to the light it was thrillingly, intensely physical. The feeling of somebody else's body touching and even grabbing your shoulder was meant to make us feel uncomfortable and to acquaint us with the idea of touching others, our future patients.

Hochstein was saying, "You need to feel that. You need to go with that. A young woman patient comes in and you're going to feel her liver, okay? You cannot think of it as an inanimate object. You're not feeling a block of wood. You know that kinesthetic sense of someone touching you, like those actors, moving you around, climbing all over you? That's what the patient feels. You've got to be in tune with them when you're doing that."

2. TAKE THE PEN

When you start in medicine, before you go out onto the ward, you have not yet made the sociologic transition to being a doctor. Consequently, there are a lot of things that go through your mind that are uncomfortable, and many of us, including myself, began to take on the patients' suffering, especially when we treated patients who were dying of cancer. It was and is not unusual for these patients to identify everything other than their disease, including us doctors, as the problem: the elevators don't work, the food is lousy, the nurses are rude. What they're angry at is the fact that they have cancer, they're going to die, and nothing they or anyone else can do will change that fact. For a young doctor, it is all too easy to take that on and go home with it, and it can destroy you.

To avoid this, we were required to meet individually and in small groups with a psychiatrist. Mine, a very prominent older therapist, was a dud, and I got very little out of our sessions. But one of my colleagues, a very empathic guy who eventually went into oncology, was more fortunate. In one of his sessions, he was sitting across the room from the shrink, articulating how much trouble he was having with his patients' anger, their problems, their complaints. The psychiatrist didn't say a thing, but he took a fancy fountain pen out of his shirt pocket and held it out in front of him at arm's length, as though presenting it. He just held it there. My friend, who was about twenty-six or twenty-seven at the time, got up from his chair, walked over to the psychiatrist, and took the pen.

The shrink said, "Did you want that pen?"

My friend said, "No, not really."

"Then why did you take it?"

"I thought I was supposed to take it."

"The patients are holding out their troubles. They are not really asking you to take them. You should only take them if you want or need to take them. Otherwise, leave it. They'll get along without your suffering. You have another job to do."

A lesser therapist might have simply interrupted, and said, "It's hard to know what's on a patient's mind, and you can't take on their troubles. You have to be able to distance yourself without being unsympathetic." All true, but without the pen, easy to ignore. The pen sold it.

3. TAKE THE WALLET

At about the same time, when I was an intern in San Francisco, we treated a farmer from the Sacramento area who had coccidioidomycosis meningitis, also known as valley fever. Stockton, California, was ground zero for the disease. It is carried by airborne dust particles that settle into the lungs, causing a pneumonia-like infection that can develop into an inflammation of the brain's lining. The treatment involves a lumbar spinal tap that allows medicine to be injected into the spinal canal daily, a very invasive regimen requiring a long stay in the hospital.

As serious as this was, the farmer had a more pressing problem. His wife was in the early stages of dementia, and she was alone on their farm with their two German shepherds. He desperately needed the treatment, and she desperately needed him.

Every day, when the team dropped by on rounds, the senior resident, whose name was Chin, told the farmer he was doing better, that his lungs were getting better, and every day the farmer would reply by asking, "Can I go home today?" The first time this happened, Chin said: "You have to understand that you have a serious infection, and you have to get antibiotic injections for the next two weeks." On rounds the following day, when the farmer again asked his one and only question: "Can I go home today?" Chin went through it all again, but less patiently: "Two weeks." After several days of this routine we were all frustrated, but Chin, who saw the world and all the people in it in the most literal terms, was beside himself. It became a problem.

One of the residents on the team, name of Kravitz, was a bit of an amateur magician and also a talented pickpocket, undoubtedly the product of a misspent youth. Tricking people means knowing and controlling what's on their minds. That was Kravitz's real talent. One day, after yet another predictable Q&A with the farmer, and during the ensuing Pavlovian decompensation by poor old Chin, Kravitz expertly snuck up from behind and lifted Chin's wallet in full view of all of us. He then suggested that we go down to lunch. He knew that Chin would offer to pay for everyone because we were interns, because he was the senior resident, because he was a very generous guy, and because Chin always paid. In the cafeteria, when he went to reach into his pocket, Chin said, "Oh my God! Where's my wallet?" As he frantically rifled through his other pockets, we saw him run through a mental inventory. "I was in the lab, then I went to pathology. I was in the chairman's office, and now I don't have it." He started to flip out. The rest of us were by now seated at a table, everyone but Chin, who kept on about the wallet.

Kravitz, who had forked over the money to get us past the cashier, changed the subject. "Let's talk about the clotting cascade," he said, meaning a diagrammable sequence of reactions in the process of blood coagulation. As he started to draw it out on a napkin, Chin, whose job was to lead us through such lessons, said, "I can't concentrate on that! How am I supposed to concentrate on that? I can't find my wallet!"

Kravitz calmly stood up and said, "Here's your wallet. Now you know what that farmer feels like. He's got a demented wife at home roaming the farm half-dressed. No one's feeding the dogs. No one's feeding her. Do you expect him to focus on all this bullshit that we're telling him when all he cares about is getting back to his wife?" Chin checked inside the wallet. We weren't sure whether he was going to thank Kravitz or kill him. Reading his mind, Kravitz smiled and said, "Oh, and by the way, thanks for lunch."

4. NEVER SHOOT A SINGING BIRD

She was young (so was I), she was strikingly beautiful even among nineteen-year-olds (I was neither), and she was seriously ill. It was the ebbing of the Age of Aquarius, and I was working the night shift in the emergency room at the University of California Medical Center off Parnassus Avenue in the post-hippie San Francisco of the mid-1970s—a vibrant, exciting place to be, even if it was somewhat hungover from the '60s. All was not well in the city by the bay, and from our perch overlooking Golden Gate Park, my fellow residents and I had front-row seats for the fallout.

Her name was Danielle, her boyfriend's name was unimportant, and we assumed that both of them had been taking drugs. When they arrived late one summer night, I noted how disheveled they both were, how typical of the city at that time. The boyfriend told me, with a notable lack of urgency, that she had had several vigorous epileptic seizures on the street after a party. He professed to love her, but could not tell me where she came from, who her parents were, or how I could get in touch with them. He seemed fairly nonchalant about the whole affair, in keeping with the tenor of the times.

Before I could get her into an examining room she lost consciousness. Her pupils were normal and she was breathing quickly. We did the usual workup—blood and urine tests for drug screens. As I was walking out of the room she was overtaken with grand mal convulsions that continued for a good two minutes. She then began to turn blue despite the administration of oxygen, and her blood pressure shot through the roof. She broke two teeth and wet herself. We immediately brought her up to the intensive care unit where she continued to have seizures intermittently without waking up. In other words, she was in *status epilepticus.*

Over the next six hours, I made desperate attempts to stop the seizures with all of the usual medications. We managed to bring her downstairs for a CT scan. It showed no abnormality, no evidence of the kind of clots in the veins overlying her brain that, in a woman her age, might have brought about the seizures. It was nearing 5:00 a.m. I had been with her through the night. The sun was infusing colors into the sky over the eastern edge of Parnassus Avenue. Out of the depths of a partially formed instinct, and even though she did not look pregnant, I called the pharmacy and asked them to send up a drip with magnesium, the tried-and-true treatment for eclampsia. We started the drip and, miraculously, the seizures stopped.

Eclampsia, a problem associated with pregnancy, causes a spectrum of symptoms similar to hers, but I hadn't even thought to check whether she was pregnant, so I ordered the test. I assumed she had been with the boyfriend for a while, that they were laid back about everything, including birth control, that they lived under the protective aura of cluelessness: "Everything's cool, everything will be okay." But I had a feeling it would not be.

When the chairman of medicine came by to make rounds with us later that morning, I told him the whole story and said that I thought her urine pregnancy test would come back positive. While we were standing there, I got a phone call from the lab informing me that it was negative. I asked him what I should do, and he said, "Never shoot a singing bird. Keep the magnesium going." We did not have a diagnosis, I had no more hunches, but it seemed to help. So I let the bird sing.

The seizures stopped but Danielle never woke up. On the following morning she died, not from the seizures, but from a separate problem of bacterial septicemia. There was no next of kin that we could track down, and the boyfriend drifted off into the dawn. Curiously, the medical examiner turned down the case, so we never found out what caused the seizures. Even though I lost my patient, the chairman's remark has never left me. We frequently do not know the cause of a problem, but we sometimes back into a good treatment. When that happens, stop being a scientist, and just keep going.

5. EVERYBODY MUST GET STONED

A year later, while on a rotation down at San Francisco General Hospital, I got a call from Bob Johnson, one of my co-residents up at UCSF. He was worried sick. He said that his sister's roommate had just called him, all frantic and disorganized, saying: "I can't wake your sister up. What should I do?"

They were living in Haight-Ashbury, so the first thing that came to my mind was drugs. But since it was Bob's sister, and because he came from an extremely respectable Wichita family—big Republican donors, five generations in the military, Bob even had a crewcut—and because Bob himself seemed to rule it out, I immediately thought, *It can't be drugs. It must be something else.*

Bob was in the middle of a shift and couldn't get away. The ambulance had just arrived at his sister's house, he told me, and they were bringing her down to the General, where I was working as the senior medical resident in the emergency room. "I just wanted to let you know, just to make sure that she's being taken care of."

As residents, we spent a third of the year at the General. The place was spectacularly but lovably squalid, like something out of a B-movie. As one of my surgical co-resident friends said when they started to renovate it, "You can't dress up shit. It's the same crappy place." The General was a converted tuberculosis sanitorium, an unholy mess, a snake pit. Parts of it had been at times taken over by a drug gang. Two female physicians had been raped in an elevator. Yet if you were in the soup and you needed something ripped open and taken care of, or if you were having a massive heart attack, it was by far the best place to go. We residents couldn't get enough of it. We loved the half-mile long halls with their giant leaded windows, loved it for the free midnight meals of tripe and Mexican chocolate with a side of cheese strata (probably just leftovers from the cafeteria that were reheated in an X-ray machine), loved it for the autonomy. We thrived on the exhilaration and novelty of serving an underserved population: the Mexicans, Filipinos, Cambodians, and Vietnamese in the neighborhood, along with the out-of-control, underprivileged, underbelly of San Francisco in the '70s. Of course, not all of our patients were victims of society.

"Okay, Bob," I said. "I got it. Don't worry. I'll take care of it, and I'll call you." Because there were no cell phones back then, this kind of give-and-take involved a lot of running back and forth to the nurses' station.

With a few charismatic exceptions, you couldn't find an attending physician within five miles of San Francisco General, but some of the best staff in the world were there: name-brand people. Because the place was, in effect, independently run by the residents, the chief resident was king and the senior resident was the prince. The most senior medical and surgical residents running the emergency room were the Tsar and Tsarina (no spirit of democracy there). As the senior resident in the ER, I was the Tsar of Potrero Hill, enjoying a brief but eventful reign.

I remember a night when a woman came in and I thought she was in Addisonian shock—that she had Addison's disease, a disorder of the adrenal glands that can lead to coma. How I sussed that out I can't recall. It was about 10:00 at night. Unsure what to do, I called the hospital switchboard to get the phone number of the head of endocrinology, who was the only full-time endocrinologist on the staff. The operator assured me: "Sure, it's okay to call him at home. He's your colleague." When he answered and listened to my plea, he said, "Are you fucking kidding me?! Take care of it!" and he hung up. So I took care of it. That was how it worked.

Bob's sister was a Midwestern belle, blonde, very attractive, a little on the plump side, but very lovely. She arrived with the usual screeching fanfare that was the leitmotif of the San Francisco ambulance corps. We had it on good faith that anyone who couldn't make it in the sanitation service was hired as an ambulance driver. There were no EMTs back then. The ambulance drivers could perform CPR and they could put in an IV. That was about it. And they could drive like mad.

The ambulance pulled into a bay surmounted by a portico, columns, and a little cupola—one of the entrances to the old TB sanitorium. Two ambulances could fit there at one time, and one bay was already taken. I watched as the attendants wheeled the girl out. She was deeply unconscious. In a car following close behind was the roommate, who rushed up to me and said, "You know, she's got diabetes." That was interesting, and would even have been useful had it been true.

They brought her to the trauma room. She looked bad: pale, sweaty, and her fingertips and toes were blue. Clearly, she was in shock and was vaso-constricting. Everybody on the team was nervous. After all she was a doctor's sister.

Nurses in trauma rooms are brilliant automatons. They're adept at shearing off clothes in one fluid motion from bottom to top. They stripped down Bob's sister, put leads on her, and right off the bat we couldn't get a blood pressure. She was about to die. They slapped on a cardiogram and it showed a very slow sinus rhythm. Tony Cimaranno was my surgical resident. He was a short, square Mafioso-looking guy from New Jersey, with the thickest, blackest, greasiest hair this side of the Rockies, and we were great together. As he used to say, he did the big cutting and I did the big thinking. "Al," he'd bark at me in a deep gruff voice, "what do I do now? I can't think for myself."

It was all hands on deck. Her blood pressure was so low that Tony couldn't find a distended vein to put in an IV. I started to do a cut down on her ankle, once a very popular and macho thing to do. There's one big vein in the medial ankle, and if you can find it and slip your finger under it, it will fill with blood and you can stick whatever you want into it. Just as Tony said, "You're never going to find it down there," I found it, put my finger under it, incised it, and slipped a feeding tube into it—a quarter-inch pediatric feeding tube for infusing large volumes of fluid for shock and trauma. Once the line was in, the nurses started furiously squeezing the IV bags to flood the body with fluid and get her blood pressure up. We grabbed some neosynephrine and piggybacked it onto the saline to give it some heft. Neosynephrine is a drug that further constricts the peripheral blood vessels in order to ramp up the blood pressure. You sacrifice perfusion to the limbs in order to keep the brain, the heart, the viscera, and the lungs getting adequate blood. The fluid was now pouring in, the neo was pulled along with it, and when I say pouring, I mean pouring. You could look at the buret hanging from the IV pole—a long transparent fluid-filled cylinder with marks on the side—and it was like Niagara Falls. Her blood pressure came up but her limbs looked awful. She was intubated by this time, and somebody was breathing for her by squeezing a bag connected to the tube. I told one of the nurses that this was Bob's sister, and she said, "What could it be? Diabetic coma? Shock?"

"She's having her period," another nurse said, "and there's a tampon in there. It looks like toxic shock."

At that time toxic shock syndrome was just being recognized. It was named in 1978, but we had been seeing cases for a few years before that, and at first had called it staph sepsis. Several of our clinicians were already onto the idea that it was due to an infection caused by tampons, and that's what this looked like at first. She had a rash, which fit, and I said, "Holy moly! You're probably right." So we grabbed some antibiotics, gave them to her, and she stabilized, but too easily. "It's not toxic shock," I said after a few minutes. We had already had a few deaths from toxic shock, I had seen it, and this definitely wasn't it.

We were now twenty minutes into it, and had been through one wrong diagnosis and one feint to the side (the roommate trying to juke us out with the diabetes). I said, "What else could it be? Let's get blood cultures and bring her up to the ICU." She was still on a ventilator. Her blood pressure was acceptable, but it was dependent on the continued infusion of massive amounts of fluids. Every time we tried to reduce any of it, the bottom fell out.

At San Francisco General in those days, the first, second, and third diagnoses in a case of coma was overdose. That was just common sense. The neighborhood was awash with drugs. Opiate overdoses were a dime a dozen. At that time Seconal was freely available on the corner opposite the hospital. You often could not figure out what kind of junk was out on the street on any given day, but if you supported the victim's breathing, they would get through it. Often, it was a cocktail of drugs that produced mental depression, excitation, and psychosis all together, as if everybody was in a rush to finish an advanced psych course and meet their maker at the same time. But not Bob's sister, no way. That just seemed too weird.

I called Bob and said, "I think you better get over here. I don't know what the hell's going on. Your sister's here, she's in really bad shape, I'm calling in Hibbard. I want him to come down and help us out." Hibbard Williams was the legendary chief of medicine.

The powerhouse service at the hospital—the medical intensive care unit—was called CHEST because the pulmonary specialists ran it. It was a coveted rotation because as a resident-in-training you were right in the meatball stew. It was like a MASH unit. Ambulance drivers would routinely go the extra five miles if they had somebody in really bad shape, somebody who needed the Navy Seals of acute medicine. A few things were probably done wrong, but more lives were saved there than in any other place I've worked. Hibbard was supposedly on his way, the team was still trying to stabilize her, and now Bob was coming. That's when our intern showed up: Linzner, a kid who was perpetually in my doghouse. Linzner came over, lifted up her eyelids, and said, "Another heroin overdose?"

I said, "What the hell are you talking about?"

He said, "Look at her pupils! Sheesh!"

Nobody had noticed. Her state of shock and the neosynephrine had given her enlarged pupils down in the ER, but now, almost two hours into it, I could see her tiny pinpoint pupils. We gave her some Narcan and she woke up almost immediately. She and her roommate had both been doing heroin.

Narcan, or naxolone, is an antidote for an opiate overdose. It competes with narcotics to bind to the brain's opiate receptors, and kicks them off so that they lose their effect. We infused it, and she woke up in thirty seconds. Her blood pressure came back. Great news, but I was fuming.

Linzner! Why of all people did it have to be Linzner?

I was constantly riding him while the other senior residents were cutting him slack because he had been sick for most of the previous year. As far as I was concerned, he was one of the worst interns we'd ever had. He was a distracted, unkempt, thick-glasses, nebbishy kind of a guy, which would have been fine if he had just done the damned work.

Yet out of the mouths of babes . . .

If anyone who was not one of Bob's friends—somebody from another service, maybe, or even Linzner—had met her at the ambulance bay, he would immediately have said, "Overdose!" He would have given her Narcan within the first two minutes, and restored her blood pressure immediately. What I should have known, what I now know too well, is that on any given day anyone can poison herself, accidentally or not, and have a seizure or fall into a coma. Suspecting an overdose of some kind is not a transcendent judgment about them or their lifestyle or their character. It is one moment in somebody's life. People poison themselves all of the time.

In spite of me, Bob's sister came through it just fine, and was none the worse. But she had taken such a massive dose that Linzner had to stay up all night to give her more Narcan every ninety minutes. Served him right.

Fortunately, Hibbard, the Big Chief, never showed up.

6. A WISE OLD BIRD

As a fourth-year medical student, my mind was set on cardiology. I had gone to the National Institute of Health to do a cardiology rotation at a time when that specialty was making a change from the classical approach—listening to the heart—to a technologically driven skill set, with echocardiography and coronary angiography at its core. As soon as I realized this was the case, I started to think cardiology didn't need me. I wasn't convinced I would be more useful than any other schmo. Then something happened that left no doubt that this was indeed true.

Her name was Denise Arduzzi, a twenty-two-year-old Portuguese-American woman who had lupus. She was in the ICU recovering from an attack when she became comatose. I can still picture everybody arching over the bed like a siege of herons: the kidney doctors who were taking care of the renal failure and the dialysis, the cardiologists who were taking care of her pericarditis (frequent in lupus), and the rheumatologists, who are supposed to be in charge of lupus in general, all of them struggling to keep her alive but unable to figure out what was going on. Then an old-time neurologist, a wise owl by the name of Robert Layzer, showed up as the consultant, and the herons made way for him. He touched, he tapped, he turned her head from side to side, and when he was through, he looked up and said, "Basilar artery thrombosis." This was before MRIs, and just as CT scans were becoming available, and it absolutely blew me away. How the hell did he do that?

"Give her heparin now," he said before he left. "Don't wait for an angiogram. Do that when you can, but give her the heparin now."

Because there was so much at stake, they gave her the heparin, an anticoagulant, and she got better. Eventually they did the angiogram, and—what do you know?—she had a thrombosis in the basilar artery: a stroke.

Who was that masked man? Before he had arrived, it was an instance of the blind men and the elephant, and not even an elephant, more like a pachyderm with humps, a mane, and horns. Then the owl drifted in, and using only his clinical skills, he cracked the case. It was a save, but it wasn't the save that sold me. If she had died it would have been upsetting, but it wouldn't have discouraged me because I was still thinking: *Where am I needed?* The herons were busy keeping her alive by giving her dialysis, but that was nothing, a trivial and almost axiomatic thing.

Sometime later I caught up to Dr. Layzer to tell him how impressed I was, and he said, "Well, that's what neurologists do." In Denise Arduzzi's case, what he did was to see what everyone else had missed: that her eyes rested downward and did not move when he rotated her head to one side and then to the other. He deduced from the hardwiring of the eye movement pathways that this must be a problem in the upper- and mid–brain stem and, ergo, a blockage of the artery that supplies it with blood.

That's when I said to myself, "This is for me."

7. THE BRAIN MAKES POETRY

In those days, medical students tended to be down on neurology. They called it a masturbatory specialty. You couldn't do much of anything in the way of cure or even prevention, so you just sent patients back out on the street ("treat-'em-and-street-'em," is how they put it). Everybody in my internal medicine residency program was appalled when I announced my intention to go into neurology. "What a waste of time! There's nothing that you can treat." Then I went to a talk given by my future mentor, who happened to be visiting the hospital. Raymond Adams, one of the greatest clinical neurologists of the twentieth century, gave an overview of where neurology stood in medicine, and, more importantly, where it was headed. I can remember his advice to this day:

Think about it: There are more diseases of the muscle, just the *muscle,* which is a tiny sliver of neurology, than there are all diseases of the lungs. There are more diseases of the spinal cord than there are of all diseases of the heart. There are more diseases of the white matter, not the whole brain, but just the *white matter,* than there are of all the rheumatologic and joint diseases. Neurology has more diseases and more complexity and more need for exquisite clinical analysis than any other branch of medicine. Take one little piece of neurology: there is more to it than in the totality of any other medical specialty. And yet lung disease and heart disease dominate our collective fears. Dermatology is the only field that has more diseases than neurology. There are a billion skin diseases, but only ten of them occur with any regularity. Whereas in neurology people may say, "Oh, it's just a stroke," well, strokes are a big proportion of neurological cases, but there's so much else to it: What kind of stroke? Where did it occur? What caused it? To master all that is a lifetime effort. So if you really want to be challenged instead of being bored in seven years, when you know all there is to know in your specialty, you may wish to consider neurology.

The brain-mind connection attracts many people to neurology initially, but that's too facile a motivation. When I hear a student say, "I want to understand the mind," I suggest they try psychoanalysis. Neurology is much bigger. Neurology isn't a system that was invented to understand the mind, like psychodynamics. It's the goddamned brain in all its messy glory. As another wise neurologist told me early on, "You want to be a nephrologist or a urologist? C'mon! The kidney? It makes urine! Who gives a shit? Now the brain—the brain makes poetry."

8. MINDFULNESS

His name was Mack. He was a tall, wiry, gentleman in his late seventies, his face covered in seborrheic keratoses, those brownish, benign, raised, flat, dry spots that older people get. Even so, he looked quite distinguished, with almost a military bearing. He had shown up at the emergency room complaining of frequent falls, slurred speech, and repetitive muscle spasms that caused him to twist his arms and torso. That triad of symptoms is particularly hard to figure out. None of the emergency room people had a bead on how to start the case, so it was appropriate to admit him to the neurology ward, to watch him, to have several people examine him, and do some tests.

That was thirty years ago. I was four years out of my residency, working as a junior faculty member at Massachusetts General, leading a team of residents. We looked in on Mack and I talked to him at some length. He was not a military man, as it turned out, but owned a package store on the South Shore. He was a Rotarian, an avid golfer, and like me a ham radio buff. When I asked him to get out of the bed, he moved slowly, holding his entire body like a block of wood while he stood at attention. That was telling. I turned to Nancy, my senior resident, and said, "I've seen this before, or something like it. He's almost too formal in his posture. He has an axial dystonia. It looks like Parkinson's, but instead of leaning forward, he's unnaturally ramrod straight. See how he holds his head back? I think this man has progressive supranuclear palsy."

Nancy, who was a very sharp cookie, turned to me and said, "How did you know that?" She was thinking that he did not have the cardinal sign of that disease: the loss of ability to look up and down.

I shrugged her off, convinced of the diagnosis, while wondering the same thing, *How* did *I know that?* I had to pause and remind myself that I was a professor, maybe only an *assistant* professor, but a professor nonetheless, and that I was rounding with residents who were five to eight years behind me. They were struggling as I once had, but it was as though I had only just now looked back and noticed that I was miles ahead of them.

Every doctor can recall a transcendent moment of arrival, a moment of breaking through the cloud cover, leaving behind the shaky confidence and self-doubt that hung over their residency and fellowship. In neurology, it takes four, five, sometimes seven years to make that transition, and once you make it, once you hit your cruising altitude, those below you in the fog might think you are performing diagnostic miracles. It is important to point out to them that you are not.

While I was still rounding with Nancy, a young woman, from a religious Jewish family, showed up at Mass General confused and sleepy, unable to feel anything on the left side of her body. This had been going on for four or five weeks. I liked her immediately. After her examination, I took Nancy aside and said, "She has a glioma in her thalamus," in other words, a brain tumor. "This one's a heartbreaker."

"And how do you know *that?*" Nancy asked.

"Mindfulness," I said.

Raymond Adams, who had hired me at Mass General, once examined a somewhat dissipated man with alcoholism and recurrent episodes of coma, someone who had recently developed speech abnormalities and clumsiness. He said, "Well, this has got to be hepatolenticular degeneration. But you know, when Kinnier Wilson described the condition in his seminal paper of 1912, he never really established the essential problem in the basal ganglia. Maybe we can get to it right now through this patient. What about these eye movements? What do you make of the way the man is standing? What about his speech pattern?"

He called the ward our laboratory, and said that it put us in a position to study the brain in a way that no other scientist could. I was struck by how actively he would listen to each patient's account of a problem, how he would quite often say, "That *is* interesting!" not out of feigned interest, but of genuine fascination. What he was thinking, he would often tell me, was that there is no laboratory or animal model for the human brain beyond this. Here is an experiment of nature: What can this patient tell us about the human nervous system?

A middle-aged woman who suffered a complete loss of sensation on one side of her body due to a small thalamic stroke came to Mass General complaining of stabbing, burning, and aching pain on that same side. Adams wanted to understand how the thalamus—a small, central switchboard for the sensory system—works, and how it could produce this pain. He started by asking a series of questions that were quite imaginative, including, "Does sexual activity alter the pain?" He had a good reason for asking. He wanted to know how the limbic system influenced the functioning of the thalamus. The question caught me off guard, but it was brilliant. How else could you figure out how the nervous system works than by querying the nervous system itself? For Adams, there was always a bigger problem to be solved, both for the brain and for the patient. That was part of his mindfulness.

Not everybody has the physical talent to be a world-class tennis player or even a good club tennis player, to take just one example, but most intelligent doctors do have the capacity to be great diagnosticians. Yet if their training and subsequent work in the field do not continually excite them intellectually, and if they have no inclination to question the experiments of nature unfolding right in front of them, they won't get there. Aptitude isn't enough. To become a good clinical neurologist, you have to be intensely interested by what the brain does, how it works, how it breaks down. Raymond Adams would walk into a patient's room thinking, *What am I going to learn now?* Because of that, he was able to synthesize hundreds and hundreds of cases, recall their essential features, and put them into diagnostic categories. Eventually, so was I.

That was the full answer to Nancy's question, but I let it go at "mindfulness."

9. NEXT PATIENT, PLEASE

In 1991 I left Mass General to expand a neurology department at St. Elizabeth's Hospital in Boston. One day, a Chinese woman in her fifties was brought to the emergency room. She had been in a coma for a week, cared for by her husband, who thought that with herbal remedies he could get her to wake up. His name was Mr. Lau, and he spoke absolutely no English. His wife's condition was critical. She had small pupils that would not react to light, she had Babinski signs (a worrisome toe reflex), and a body temperature of only 91 degrees. I added it up: she was Chinese, had hypothermia, had hydrocephalus, and despite her normal chest X-ray, I decided she must have tuberculous meningitis, with tuberculosis somewhere other than her lungs, and almost certainly brought over from China. She needed to have her ventricles drained, just like Mrs. G.

Mrs. Lau's case had lain dormant in my mind until the resuscitation of Sofia Gyftopoulos, the woman whose code blue had caused me to miss a Red Sox game. In the aftermath of that code blue, Hannah had walked into my office for a debriefing. She was more deflated than distraught.

"The bottom line," she said, "is that you saved her life. If you had not been there, she wouldn't have gotten the mannitol, wouldn't have gotten the drain, she probably wouldn't have had the first shock in time, and she probably wouldn't have made it."

"And you're wondering what you would have done?"

"I know what I would have done. Nothing."

"That's not true. You would have figured it out."

"But not in time. It was humbling, almost to the point that I felt like I'm never ever going to be there. It made me almost think I might as well give up, because I can't even imagine a day when that would happen, when I would walk into that room, and I would say this is what needs to happen, and I would be able to make it happen, and that person would end up living because of me."

I told her about Mrs. Lau, about having seen a similar constellation of signs before. I told her that although I was right about that case— that it was tuberculosis, that it had spread and caused chronic meningitis—Mrs. Lau died despite all of our efforts, and despite draining her ventricles. My diagnosis was confirmed at autopsy. But it was because of that experience and others like it, because I had witnessed unusual, extraordinary cases, that I knew what was going on in Mrs. G's head when she went into cardiac arrest.

"You're not going to save everyone," I told Hannah, "you're not going to figure out everything. It's part of the process. What you need to focus on is the far greater proportion of people you have pulled out of the fire, that you will pull out of the fire. Then you move on to the next case. Right now, you're just trying to stay one step ahead, but don't lose the opportunity to appreciate what the signs and symptoms are teaching you about the brain. It's to the patient's benefit if you soak up the details, if you're really into it. The farther away you get from the little details, from the fascination with what the brain does, the less of a clinician you will be. So just keep moving in that direction. Mrs. G was your Mrs. Lau. Now you know what to do."

8

Endgame

Facing down Lou Gehrig's disease

Louise Nagle had been kicking around doctors' offices and emergency rooms for three months, complaining of breathing trouble while climbing stairs. She was referred to pulmonary and cardiac experts, who could not sort out the cause. Understandably frustrated, she came into the Emergency Department one night at about 8:00 after almost choking to death on a dinner of cubed chicken, peas, and rice.

Hannah was the senior resident on duty. When we entered Mrs. Nagle's cubicle, we saw the trouble almost immediately: a flutter that can make your heart sink, a slight palpitation between the thumb and first finger that signifies degeneration of the motor nerves. In other words, ALS, or Lou Gehrig's disease.

As Hannah and I stood over her bed, we could feel Mrs. Nagle's distress. Her nostrils flaring like a racehorse's, she was using every muscle in her shoulders and thorax just to breathe. Beads of sweat welled up on her forehead and cheeks. She looked vaguely like someone having an asthma attack, but as she grabbed the side rail of the stretcher with her left hand in an attempt to sit up, I kept my eye on her right hand. It lay motionless on the sheet. If only it *had* been motionless. I don't know why I looked, but I couldn't help noticing the butterfly of doom flitting between her thumb and finger.

Fasciculations—the fluttering of small parts of a muscle, or what used to be called live flesh—are almost always benign. Virtually everyone has had the experience: small contractions around the eyes or the mouth, in the calf or the forearm. It happens when people are tired or they've had too much coffee or alcohol. Tending as they do to come in clusters over several days, these muscle flutters can reflect a patient's self-involvement or their fear of illness. When that patient is a medical student or a physician, he (it's usually a he) often becomes immobilized by the experience, and will either demand an immediate medical examination or simply hide in fear, hoping to cancel out a degenerative spinal condition by ignoring it. Although my evidence is entirely anecdotal, I have become convinced of two things: (1) physicians complain disproportionately of this symptom and are hard to persuade that it is benign; and (2) only the nicest people get ALS (thus immunizing a substantial number of physicians).

Benign fasciculations can lead to a cramp, but are unassociated with muscle weakness. They go away. The bad twitches, such as the one I saw in Mrs. Nagle's hand, result from the dying of nerve cells in the spinal cord. They occur when a group of muscle fibers loses the connection to its controlling nerve cell—the dying cell—and an adjacent surviving cell then takes over control of those fibers. This new, larger, motor unit cannot easily be sustained by the single nerve cell, and becomes very unstable, producing the spontaneous contractions in the remolded motor unit.

Mrs. Nagle's case was advanced enough, as evidenced by the muscle flutters and by her inability to easily move air in and out of her chest, that she herself would soon cease to be a fully functioning organism. That's one way of putting it. I had to close my eyes for a moment and let her think that I was just tired, but seeing her husband

and two young children chilled me. Hannah and I performed a perfunctory neurological exam and affirmed all of the core features of ALS. The flickering muscles in her hand were shrunken, atrophic. There were fasciculations in her other hand. Her reflexes were all brisk, too jumpy for an otherwise healthy middle-aged woman, a sign that the nerve cells traveling from her brain's motor cortex down to the spinal cord were also dying. It took a split second for me to suspect what was going on, and no more than two minutes to confirm it. Out of the corner of my eye I noticed that there was a multitude of other muscles that were twitching the same way. That, combined with her too-brisk reflexes, signaled damaged and dying motor nerves.

There is a constant physical and emotional distance that separates a doctor and a sick patient, as though a rising tide were filling a channel that lay between them. But that space is not entirely empty. It is filled with emotion—hers of anguish, mine of a welling sense of dread. As long as my dread remains hidden, I can muster the degree of calmness and patience that allows me to stay at the bedside and be effective. Knowing how high the tide will go, I can sometimes see that it will take my patient with it. At such moments I have the comfort of knowing that I will be going home to my wife, my kids, and my dog, with all of my faculties intact. With Mrs. Nagle, I had to remind myself that I could only be useful to this person by disconnecting myself from the pain to come. This willful mental posturing affords me the degree of internal quietness I need to stay at her side and let her fear fill all of the space separating us. It is a feeling that comes with some power, I suppose; it is a feeling that allows me to stake out a moment of existential peace for both of us.

How and when to tell her and her husband? Hannah gave me a look that said, "We have to tell them sometime." I gave her one that said, "Do they really need to know all of this tonight?" There was no question of sending her home, where she would probably have a respiratory arrest or drown in her own secretions. But telling her that she had a progressive illness, which would be fatal in months unless she committed to life on a ventilator, and if she did not, would likely entail an ugly death by gradual strangulation—breathlessness, air hunger, and emaciation—I decided that this was not an appropriate midnight discussion, and waved off Hannah's signal.

Yet they had already been told what she did *not* have. It was not a lung problem, not a heart problem, and not asthma. It had been suggested that she was simply overanxious and overwhelmed dealing with two young children at home, and that she was probably hyperventilating. That would have been reassuring, but she now knew that she had something more serious than any of these comparatively desirable alternatives. There was nothing to be gained by shocking her further. I told her that this was likely to be a neurological problem, and that, at a minimum, we would be able to give it a name, that it was something we had substantial experience with, and that it could be managed. In other words, despite my best efforts, I said too much.

"Is it Lou Gehrig's disease?"

"Well . . . that is certainly a possibility. Why do you bring that particular disease up?"

"I looked on the Web and it said that Lou Gehrig's disease could cause muscle weakness, twitching, and sometimes make it hard to breathe."

"The Web is usually wrong. Why don't we wait and see."

"But what if it is Lou Gehrig's? What's going to happen?"

"There are many ways things can go, so why don't we just make you more comfortable tonight until we can sort it out."

To my relief, the act of talking so exhausted her that she ended the conversation. How did she know? It reminded me of a ridiculous joke I'd heard on *The Sopranos*: "What are the chances that someone named Lou Gehrig would die of Lou Gehrig's disease?" My next thought was that she was probably a wonderful person, and that this diagnosis rather cruelly proved it.

The following morning on rounds, as I stood facing her from the right side of the bed, I saw that she was breathing more comfortably.

One of the residents had given her a positive pressure mask typically used for sleep apnea. It reduced the work she had to do to breathe, and it let her get some sleep for the first time in several nights. She was exhausted nonetheless, and had the look of someone who could not decide whether she wanted to put up a fight. With the low, early morning light from Francis Street running diagonally across her body, it was now evident that virtually every muscle in her body was twitching away like an orchestra of fleas—"widespread fasciculations," as we say. Despite this, her strength was reasonably preserved, which, unfortunately, only reinforced the diagnosis.

For two days I vacillated. Before I went to her with something akin to a legal verdict of ALS, I wanted an electromyelogram test to confirm the diagnosis. This was largely to reassure myself. Once I committed to it, it would be a very difficult diagnosis to withdraw. Such is the case with all fatal diseases: if the bad news you've delivered turns out to be incorrect, in my experience, the reprieve provokes a great deal of anger in the patient.

When the EMG came back I could read it in black and white: "Widespread fibrillation and fasciculation potentials including all four limbs and proximal and paraspinal muscles." As Bette Davis's neurosurgeon put it so succinctly in *Dark Victory*: "prognosis negative."

Louise Nagle was an only child, the daughter of a prominent physicist. Born and schooled in one of Boston's tonier suburbs, she had gone to Cornell, earned a master's degree in city and regional planning, studied and traveled extensively in Europe, and had eventually returned to Boston. She had married in her midthirties, had come to accept the quotidian suburban normality, along with the dearth of the kind of wonder that she had previously experienced daily in a city like Rome. She became a self-described soccer mom, a self-assessment delivered with what I interpreted as an air of regret. Her own days as an athlete—a swimmer and a gymnast—were long gone. Her days as an artistic adventurer were, too. And now her days as a living, breathing person were running out.

"Louise, most of the evidence from your examination and EMG points to a problem called motor neuron disease."

"You mean Lou Gehrig's disease?" she whispered, more declaratively than interrogatively.

"Yes, it is a form of Lou Gehrig's disease."

"So it's over?"

This time there was a question, and it had to be answered. She wasn't stunned. She was expecting this diagnosis, but the lilt of her voice told me this was not a prepackaged reaction. It hadn't sunk in yet, and it was my job to hold her hand while it did.

"No, it is not over. Not by a long shot. There are many things we can—that we *will* do—to keep you going, keep you at home, and make you comfortable."

After assurances that we would not abandon her, and that things would be set in motion to provide the support she needed to be able to function at home and to participate in her family's life, the tension dissipated, and Act I drew to a close. We discharged her. She went home with a considerable amount of paraphernalia for breathing assistance, a device for suctioning if she began to choke, and instructions for her husband about the Heimlich maneuver.

Within a week there was a frantic phone call on my office answering machine. "Doctor, my wife is strangling on food. What should I do?"

This was the beginning of Act II.

The call came in at 6:15 a.m., and my secretary, who arrives at 8:00, would not pick it up until about 8:15. By that time Mrs. Nagle was at a local emergency room, and the doctor there called me. "She's actually fine," he said, "just very anxious and tachypneic [breathing quickly]. Her chest is clear and she has no fever or sign of aspiration. She probably just had a little trouble getting some milk down and freaked out." I asked whether he thought it appropriate to give her a bit of clonazepam for anxiety, and he agreed that it made sense.

The next week it happened again, and Mrs. Nagle went to the same emergency room. It happened two more times over the ensuing three weeks. Each time I received a call, and after her fifth ED visit, this time at the Brigham, I asked the Nagles to come see me in the clinic at the end of the day. I began to suspect that her anxiety was causing more physical discomfort than the disease.

"I simply can't breathe a lot of the time," she said. "I feel as if I'm going to choke and then die. I'm so worried when I'm alone."

"Are things getting stuck in your throat or chest, or are you just feeling as if you can't get enough air?"

"Yes, that's it. I can't get enough air. I never feel as if I can expand my chest enough."

"Which comes first, feeling anxious or the trouble breathing?"

"They come together. I can't tell which is which, but I do think that I'm going to die."

The fact that she could get a few consecutive sentences out without taking a breath indicated that her respiratory apparatus was working more than adequately, that there was no medical reason for the perception of shortness of breath. I listened to her chest with a stethoscope, and it sounded clear. She was breathing at about eight breaths a minute at rest. She accepted the notion that she had a disproportionate degree of anxiety about the current state of her illness, a level of fear pegged to the endgame rather than what was happening at the time. I decided to go out on a limb that only an experienced senior physician can climb onto.

"You are not going to die anytime soon."

We talked about being aware of the anxiety and its effects, about controlling it with breathing exercises, about meditating in the morning. Her reaction, I explained, was understandable in the circumstances, but the disease was not the main cause of her breathing difficulty, not yet anyway. I told her that I didn't think antidepressants would help at that stage. I did not mention that they might eventually come in handy. Things would get a lot worse.

Anyone diagnosed with ALS has a series of gut-wrenching decisions to make. After the shock has worn off, once the capacity for rational thought returns, many small choices and a few large ones come along in relatively rapid succession. When will I need a wheelchair? Will I accept a feeding tube? Will I go on a ventilator machine?

The pace of the disease's progression varies from person to person, but the end result does not: eventual loss of all motor function with the exception of eye movement and blinking, and possibly some lip movement. Because the tongue can go relatively early, speech does too. The fasciculating muscles begin to atrophy. The ability to swallow is soon compromised. The disease is most certainly fatal if left to run its course. Yet it leaves the brain and the sensory system intact. The patient can feel everything and move almost nothing, eventually coming an eyelash away from being completely locked in. The ultimate question facing the ALS patient is whether to be or not to be. Quite simply, will I do whatever it takes to stay alive, in whatever condition, or will I let the disease take over? In practice, the question is much more direct: Are you willing to have a tracheostomy tube inserted and be placed on a ventilator? If so, you will no longer be able to talk, but you will be able to breathe. The majority of patients opt not to do this. Initially, Louise Nagle made it very clear: no ventilator, no feeding tube. But the decisions we make in relative calm and those we make under duress are not always the same.

———

In the fall of 2002, George Kalomiris had everything to look forward to: a lovely wife, a new daughter (just three months old), a new house in Nahant with commanding views of Revere Beach and the Boston skyline, a budding career in state government. George himself had been living a dream. A basketball star in high school, he had been recruited by several Division I colleges, but chose to go to Tufts for the academics, to star on the basketball team, and to stay close to his Greek immigrant parents as their dutiful only child. After graduation, he played ball professionally in Greece for seven years. A dominating power forward at six feet four inches, he played "above the rim." Upon his return he entered local politics, and eventually landed a job as the Director of Economic Development for the Commonwealth of Massachusetts. After marrying and building his dream house, he was poised to reap the rewards of years of hard work.

George and Felicity Kalomiris are not quite sure when the first signs of ALS appeared. George thinks it might have been in early 2000 when he went quail hunting with some friends in Rhode Island, and found his shotgun to be unusually heavy. Or it might have been when he was changing the baby and had trouble with the pull tabs on the disposable diapers. At the time, Felicity thought he was just trying to shirk his duty.

He can recall the twitches, but everyone has twitches, and these seemed to correspond to a recent return to the weight room after a long hiatus. Nothing unusual there: just work through it, he thought. But two months later he sprained his thumb merely by knocking it against a table. It wouldn't heal. When a doctor finally looked at it, he noticed atrophy in the surrounding muscle tissue. Only then did George notice it himself: hollows and indentations where there had once been pads of muscle. That's when he came to see me.

George Kalomiris considers himself lucky. He is not in denial. He has good reasons to think so, and these become clearer when you meet him. He is still alive a decade after his ALS diagnosis. His body may be confined to a wheelchair, but his mind is not. George had planned wisely. He had signed up for long-term disability insurance when Felicity became pregnant. He had gone to a financial adviser who gave him the kind of advice that is given to all new parents, but unlike most new parents, George took it. With his own insurance plus Medicare, he could afford to pay a rotating team of at-home caregivers. His disease, compared to most (Louise Nagle's, for instance) was of the slow-progressing variety. But none of those things mattered. They did not figure into his decision to live as long as he possibly could, in any condition that allowed him to watch his daughter grow up, that allowed him to deepen his relationship with his wife, that allowed him to sustain the friendship of the people he loved.

"There really was no decision," Felicity explained. "There was never any doubt or thought of how we would pay for it or how we would manage."

Whenever a new case of ALS comes along, I think of George and wonder how he's doing. It had been five years since our last visit, but I knew he was still active, if only on the Internet. I didn't know how his home life was going, but I decided to find out, partly for selfish reasons. We all succumb to something eventually, but few of us get to decide when to cede control of our lives. George hadn't ceded yet, and probably never would.

Fast-forward four months. After a long interlude of relative stability and calm, Louise Nagle arrived in our emergency room in the very same straits that had previously sent her to the local hospital. This time she was breathing at about twenty-nine times a minute rather than eight, she was sweaty all over, and could barely get three words out without gulping for air. This was the real thing. Her diaphragm, one of the largest muscles in the body, was running out of gas.

At her previous office visit we had discussed, but only in general terms, what measures she would want us to take if her breathing worsened and began to fail. She and her husband had previously decided that she would not go on a breathing machine or have a tube stuck down her throat. They had discussed this at my suggestion, and now came back with their final decision. It was not the one I expected, although I should have.

"I just . . . I just . . . don't want to die now."

"Louise, you look extremely uncomfortable," I said, "and I would like to put a breathing tube in because you're working so hard that you'll soon fatigue and be unable to breathe. You need to understand that you will then be on a breathing machine, and it may be difficult to get you off it. And you won't be able to talk because of the tube in your throat."

"Okay. Okay. Do it quick."

She was now angry and agitated, unable to find a comfortable position in the bed with the sheets twisted up around her. After a quick blast of the midazolam to sedate her, the tube went in easily. By the time she woke up from the drug, the ventilator was working for her and she was wide-eyed and comfortable, relieved, gripping my hand.

"Is this okay? Is this what you want for the moment?" She gave me a quick but definite nod.

Like other patients who have ALS, Louise Nagle had full command of her faculties, if not a kind of hyperawareness. Crises always focus the mind, so much so in her case that she had completely rethought the breathing tube, decided that in fact it was now quite okay because it was not an irrevocable decision. It was okay because it would afford her the opportunity to assess her situation in calmer and controlled circumstances, so that she would not be forced to make the big decisions in the midst of a crisis, or even in a theoretical framework that afforded her only the idea of dying, and not a taste of what it really felt like. I told her several times that she had control over her destiny, that we would not coerce her into a course of action or inaction, but would support her in any direction she decided to take. I was very specific that at any time she wished we could take the tube out and sedate her so that her air hunger was suppressed. I was clear that this would mean imminent death while under deep sedation. I did not back away from offering to guide her, and as our conversations progressed, I told her what other patients had done. I told her about George, my lone holdout. And whether my personal judgment of him vacillated between bravery or selfishness, I had to concede this much, that his determination gave me a perspective on the other choice an ALS patient can make, the choice that few do make, or can even bear to make—to be rather than not to be.

George never had a moment's hesitation, and it was not in him to rescind his decision. But he was very clear that he did not judge those who chose the other path. He had come to know many such people, had befriended them, and had helped them with their choice. Louise Nagle was now facing that decision. Not immediately, but soon. All ALS sufferers face a kind of Hobson's choice. They will die. The only question is when and how. How much suffering are they willing to bear? How much incapacity? How much of a burden are they willing to become?

A few days on the breathing tube stretched into two weeks, sufficient time for the presence of the tube to soften the rings of Louise's trachea, creating even more problems. The tube would have to be replaced by a tracheostomy that, while more comfortable, would still prevent her from speaking—no air would pass through her vocal cords and mouth—and it would commit her to a respirator day and night for the foreseeable future. The alternative was to take the breathing tube out under controlled circumstances while she was sedated, and allow her to die. It was time for another conversation with the whole family.

"You understand, Louise, the tracheostomy is another step toward prolonged support using a ventilator machine?"

A positive nod in response.

"It involves an operation on the front of your neck that is brief. You will be under anesthesia. Just as with the tracheal tube, this step is not irreversible. My goal is to make sure you're comfortable both physically and psychologically at every point."

Her husband's only question was: "It won't be done by a student, will it?"

"It will be done by a surgical resident under the supervision of a staff member, and that will be fine."

"So, someone who has barely done it before?"

"Not exactly," I said, with intentional ambiguity. The procedure is routine enough that it rarely requires a senior surgeon.

The resident inserted the tracheostomy the next morning. By week's end we were able to get Louise into a nearby chronic care respiratory hospital so that I could stop by periodically, and it would be a short ride to our emergency room if there was an acute problem. Things seemed calm enough when I went to visit her. She was virtually motionless in bed, but could mouth some words and signal with her eyes. I brought in a large clear Plexiglas board that had the alphabet and a few short words written on it. I showed Louise how I could see what letter or word she was looking at from the other side. Communication became a silent telegraph of terse answers projected through Plexiglas.

"Louise, are you comfortable?"

She scanned to "N" on the board. Now began the laborious process of determining the source of that "no." This was not the time to walk away and ask the nurses to figure it out.

"Are you in pain?" Again, her eyes diverted to N.

"Are you getting enough air?"

To N again.

Air hunger is one of the most uncomfortable symptoms known to man or woman. Struggling to breathe, the awareness of breathing, contemplating the difficulty of each breath—something that a healthy, resting person never has to think about—is the feeling of dying. It probably contributes to the imminent sense of death (*angor animi*) of a panic attack.

Louise had had this experience before, out of anxiety. This time it was caused by the shift from a breathing tube in her mouth to the new tracheostomy in her neck. Given the small diameter of the new tube, it must have felt like breathing through a straw. I took a quick look at the control dashboard of the ventilator. The rate and volume looked good, but a small alarm on the panel intermittently blinked LOW EXHALED VOLUME. This is one of the most vexing problems in ventilator management, because it means that air is being lost somewhere between the time it is pushed into the lungs and when it comes back out.

Most often the loss occurs on the way in, either into the chest or somewhere around the cuff that holds the tube in place in her trachea.

I grabbed the tiny bag of the air reservoir that hangs down from the tracheostomy tube like a piece of spaghetti. It was too easily compressible, indicating that the cuff around the tube wasn't inflated. Grabbing a syringe, I slammed a few extra milliliters of air into the cuff. She began to sweat all over and tremble a bit. I saw that her heart rate had gone up from around 92 bpm to 135 bpm. She was struggling. Her eyes widened and her pupils dilated. There were beads of sweat on her forehead. The additional air I injected had not expanded the cuff. I grabbed a compressible Ambu bag, disconnected the ventilator, and put the coupler of the bag right onto her tracheostomy. By squeezing the black bag slowly and firmly, I could move enough air in and out to expand her lungs at a comfortable pace. Her heart rate settled down below 100, and her eyes no longer looked as if they were going to pop out of their sockets. I heard the French horn rumble of air escaping upward through her trachea and around the leaky cuff.

"Better?"

A tentative nod. I asked the nurse to call down to the supply room for a replacement tracheostomy tube. A stunned nursing student took over the bagging. I had to tell her to do it with both hands, and smoothly. I don't like to be overly solicitous with patients because it can seem gratuitous and end up as anything but reassuring. But I took Louise's hand, damp with sweat, and held it just firmly enough to let her know she was still connected to something. The look in her eyes gave me a sinking feeling that this experience had clued her into what might be coming her way.

We slipped the new tracheostomy tube in—a messy affair, with mucus dripping out of the opening onto her neck in a frappe of blood mixed with saline, with small bubbles around her trachea. She probably wanted to cough but could not generate the contraction in her diaphragm. The nurse suctioned out the tracheostomy. More blood.

Things settled down for her, but I had missed the first half of my clinic, putting additional pressure on me to move along. At times like these, I have always felt it best to step back and let the world take care of itself for a few minutes. The idea that I could be in two places at once fell off the table for me decades ago, and I realized that, in the end, it just means being late for dinner. Over ninety minutes had passed since I first went up to see how she was doing. The contrast between her initial groan of breathlessness and now the tapping samba rhythm of the ventilator, the whooshing in and whirring out, was as comforting to me as it was to her. Respirator Zen.

We had reached the next level of her descent, the point of no returning home. I called the respiratory hospital a few times, but waited a week until I visited again, hoping that her husband would be there.

"That must have been a very uncomfortable situation, wasn't it, when the tracheostomy leaked? This kind of thing happens infrequently. We can handle it if it comes up again."

A single long blink—yes.

"Has enough time passed that we can talk about what we should do going forward? Are you okay with the way things are?" She signaled her response by looking to her extreme left where her husband stood.

"We had begun to talk about this before Louise got so sick," he said, while looking out the window. "The kids and I have been over and over it, and I just can't come to grips with what's happening. Louise wants this to be over with. She can't take being stuck in bed with people turning her from side to side and changing her diaper. This business of having to be fed through a tube and breathing through a ventilator is killing her."

"You do understand that on the ventilator she could be kept alive almost indefinitely, she could see the kids grow, and her mind would be good?"

"Yes, but she's stuck on not wanting to be a burden to us."

Her facial muscles could barely mount an expression, but she was obviously tearing up.

"It's tough for me as a physician to judge anyone's quality of life, Louise. I owe it to you to make sure you're comfortable no matter what you choose, and to make sure there's no pressure to make any decision." We had been dancing around this for almost a year, more intensely since her respiratory episode of the previous week. Was she going to ask me to ease her way out? Was I going to suggest this as a real possibility to her?

I can only assume it requires a long internal conversation to make a confident decision to die. Louise was not in physical pain, was not suffering from lack of air, and had no bedsores to produce discomfort. But no amount of reassurance from her husband that she was not a burden could possibly ring true. To arrive at a decision to die because of the mental anguish you are causing your loved ones is, to almost any of us, an absurd and anguishing hypothetical, the stuff of made-for-television movies. An Alzheimer's patient who drives his family and the nursing staff utterly crazy lacks insight to make this decision. Would Louise be selfish if she wanted to end it? Would she be magnanimous? She had latched onto the diagnosis of ALS so rapidly, despite my euphemisms, that I was inclined to think she would take the lead. But again, I looked around and found all eyes on me.

The respiratory hospital is a world unto itself, a kind of purgatory, a place where tomorrow and tomorrow creep in their petty pace. The cheerful and encouraging staff do what they can to mute the peculiar odor of sweat, feces, disinfectant, and hopelessness. Louise's husband, sitting in a chair with his back to the bed, looking out the window, shifted toward me slowly, staying seated while making the pivot. A depressive move, I thought.

"Doc, she wants it over with."

"Have you thought about how everyone would feel when you're gone, Louise?"

A nod, and this time with no hint of tears. She was clear-eyed and clearheaded, as best I could tell.

"Louise, are you depressed?" With all her energy, she took an in-breath and exceeded the chest movement caused by the ventilator. Her eyes darted around to the plexiglass board resting on a chair. I picked it up and put it at arm's length between us. She went right to the letter N.

"Do you mean you're not depressed?"

Another deep inspiration. Now, with a determination that she signaled to me by rapid and emphatic movement of her eyes from one letter to the next, she spelled out:

"N . . . O . . . I . . . A . . . M . . . N . . . O . . . T . . . D . . . E . . . P . . . R . . . E . . . S . . . D"

"Okay then."

The next day, she spelled out: IVE HAD ENOUGH. And then: NEVER BELIEVED ANYBODY COULD GET THIS SICK.

I arranged to have her transferred back to the hospital under the pretense of worsening respiratory failure. It was a complete canard since she had barely any respiration to fail. The residents were puzzled about why I had admitted an ALS patient to the acute neurology ward, but I took Joelle, the senior ICU resident, aside and filled her in. Some day she would find herself in the same position. Yes, there seemed to be nothing acutely wrong with this patient. She was here because her neurologist was going to fulfill his promise to her and her family, and see her illness through to the end.

The medical capability to keep someone going, while not limitless, is long. We could give a leukemia patient ten bone marrow transplants. We could try experimental chemotherapy. We could keep giving them platelet transfusions. A similar array of extreme measures is available to the ALS patient, but none of them changes the disease. So the question becomes: If you deny a patient *everything* that medicine has to offer, are you assisting in her death?

In a hypothetical case of assisted suicide, the patient is saying: "I don't want to live. Give me something." But in a case like Louise Nagle's, the patient is saying: "Do you have to keep treating me to the nth degree?" We don't, but when we don't, Death leans in. Before that happens, a delicate conversation has to occur.

"We have to face the reality that without treatment you will die. Do you want me to make it less difficult, even if it accelerates things?"

Any physician in a hospital setting faces this scenario all too often, and is obliged to negotiate with Death through the dying patient. And it is not uncommon for the patient, after a period of consideration and reflection, to say, "Yes. Please. Let's stop this."

What happens next is both ugly and peaceful. There is no way to delegate the act of easing someone out. All the chart orders in the world can't capture the weight of the medications that will be slowly dripped into the vein. In the modern era, I am still not free to be the agent of someone's demise. It's not in me to do the deed anyway, and it would not matter one way or the other if the next ballot question on physician-assisted suicide passes or is defeated. What Louise and her husband had decided, since she was entitled to her autonomy and was fully aware of the consequences, and according to my psychiatry colleagues she was not depressed, was that she would have the tracheostomy tube removed. That was her right. In order to prevent the horrific air hunger that would result, I would give her small, then increasing doses of morphine before and after the tube came out.

Louise Nagle's diaphragm still functioned, but not enough to keep her alive. The morphine would make her groggy, but it would also make her breathing easier. I wanted to get her through a window where she would be awake enough to speak, but wouldn't suffer. If I managed to take her to there and then removed the breathing tube, nothing would obstruct the flow of air through her larynx. She could utter a few words while her breath held out. She could say her final words to her husband.

I hadn't raised this point before, hadn't forewarned her, and by the look in her eyes I could tell that she was shocked. She hadn't prepared a speech, but then, how could anyone prepare such a speech? "You have a few breaths left," I might have said, "one more breath with which to say what you have to say before lapsing into unconsciousness forever." Too melodramatic, perhaps not fair, but the time had come to remove the tube.

I stood by the bed, my sport coat slung over the bedside table. Louise's husband sat in a chair on the other side of the bed with his hands resting on the sheets, clasped prayer style, his forehead almost touching the mattress in the triangle of his arms. I couldn't see whether his eyes were open or closed. The sheets were messy, creased, a bit sweaty. A smoky light from the ward penetrated the room and lent it an air of gloom. I could hear the people in the hall, their murmurs, their squeaky soles, their dinner trays, all in discord with the finality of death. Henry Nagle, immobilized, became part of the furniture.

Louise looked up at me, unblinking, as I nodded the go-ahead, then she blinked her consent and closed her eyes. I dialed up the morphine to make her sleepy, and I waited. Her eyelids fluttered.

"Are you awake, Louise?"

She stirred and nodded.

"Are you okay? Are you getting enough air?"

Another nod.

"Okay, what I'm going to do is take the tube out, and you may be able to speak, to whisper a few sentences. But I'm not going to allow you to become uncomfortable. If it looks like you're breathing too hard or laboring, I'm going to turn it up, and you might not have that opportunity."

She was arousable, sleepy but arousable. The ventilator indicated that she was pulling a trace of air as I disconnected it from the tube and cut the cuff reservoir. I could hear a stirring of air moving over her vocal cords. I suctioned around the tube with a catheter so that she wouldn't drown in her own secretions, and then slowly pulled the tube out.

You can tell when someone is about to die. Their vessels constrict as blood pressure starts to fall, and the kneecaps get a little bluish.

Louise made a few furtive coughs, almost like huffing. Then I heard a noise, a hiss, and she whispered, "I . . . I . . . love you. Good-bye."

Henry hadn't moved.

When I got back to the office, I said to my secretary, "Get hold of Felicity Kalomiris and set up a visit with her and George. Please."

9

The Examined Life

What it takes to survive a motor-neuron death sentence

"I had this guy come by yesterday, this twenty-nine-year-old guy who shows up in the office with his wife. They're from the Deep South. Mississippi. White guy. Cracker. A slob of a tattooed smoker."

Elliott was telling a story. It was Saturday morning, and he was telling me a what-the-hell-brings-you-here story. We had run into each other at the hospital on a mutual day off, and we were perched on tall stools at one of the café tables at the entrance to Au Bon Pain, a place which not only fuels the Brigham staff with caffeine, but also supplies the bulk of the empty calories and artery blockers. It is tucked inside the hospital's main lobby. I should have been home fixing the guy wires on the ham radio antenna on the back of my house, the bane of my neighbors' existence, but a lifeline for me. Instead I had come in to check on Hans, one of my Parkinson's patients, in order to adjust his deep brain stimulator, after which I was expected in Nahant, at George and Felicity Kalomiris's house. I hadn't seen George in five years.

Elliott was on a roll. "So the guy says, 'I'm in pain, PAIN! I can't touch here, I can't touch there! My whole body . . . it hurts, and it's just killing me.' And the wife is saying, 'Oh, baby, oh, baby, tell the doctor about that other doctor you saw.' And they hand me a paper that says 'Decrum's disease.'"

I nodded at the reference: a condition with multiple little lipomas, or benign soft tissue tumors, that can be very painful.

"She says, 'That's what he has! That's what you have, baby.' So they show up here—listen to this—they drive all the way up here from Mississippi, with no place to stay, no money, no insurance, and they tell me, 'We're staying in Boston until you solve it.' Now this is a guy with no intellectual capacity at all—a moron—and he goes on the Internet, and he finds this Decrum's disease, which of course isn't Decrum's disease at all, but *Dercum's* disease. 'I hurt all over, I've got Decrum's disease!' He's wailing like a banshee. It was the most disembodied type of nonsense. Blatant. Of course what he's got is a major psychiatric problem, as well as being a nut."

This passes for conversation sometimes. We get all kinds of patients, some of whom we help to live, some of whom we help to die, some of whom we can't help at all. All we can do is listen.

Before Elliott could get any more traction with his tale of woe, a rumpled old guy in a trench coat materialized by his right shoulder, and as if Elliott wasn't even there, as if he didn't happen to be right in the middle of a sentence (which he was), said to me:

"Somebody just pointed you out and said, 'When this man has a friend in need, *that's* a man he stays away from. Is that true?'"

"That is true," I replied automatically. "And do you know that for a fact?"

"No, but somebody pointed you out."

"And is your name Henny Youngman?" I asked him.

"I used to write for Henny when he lived on Thirty-five East Thirty-fifth Street." By the looks of him, that was entirely credible. "The first joke I wrote for him: 'I ate at his house and his wife was the world's worst cook, and I said, 'Now I know why you say your prayers *after* you eat.'"

Very good, I thought, even though it's a Myron Cohen joke, but good enough that I returned the volley. I couldn't help myself.

"My secretary just buzzed me and said, 'Doctor, there's a man out here who thinks he's invisible.' 'Tell him I can't see him.'"

Elliott gave me a what-am-I-chopped-liver look, but our new friend ignored him and pressed on.

"I just bought my girlfriend a diamond ring, and she says, 'Gee, thank you. It reminds me of Arkansas.' I say, 'Why Arkansas?' She says, 'Little Rock.' I have another girlfriend, her name is Bubbles, and Bubbles was wearing very tight skirts. They were so tight that I became concerned about her circulation. Then I realized, the tighter the skirts, the more she circulates."

"You can't be a local. Are you from New York?" I asked him.

"I live in Framingham. You know what happened, I got a call from one of your colleagues this morning, he said, 'The check you sent me came back.' I responded, 'So what? So did my arthritis.' My marriages were so horrible, my first marriage, my wife divorced me, my second marriage, my wife won't."

The guy was a throwback to a time and place I identify with.

"What's your name?"

"Sheldon. You call me, we'll have some fun."

"Dr. Allan Ropper," I said, extending my hand. "It's been a pressure."

He gave me his number, but like any good comic, he had to have the last laugh.

"You know what happened to me this morning? At three a.m. my neighbors began pounding on my door, POUNDING on my door— would you believe? At three a.m.!? Luckily I was still awake playing my bagpipes. That raises chutzpah to new heights." And he was gone.

"And this is not unusual for you?" Elliott said.

"It's a curse."

Felicity and George Kalomiris live in Nahant, a peninsula that juts out into Massachusetts Bay some ten miles north of Boston. The views from their living and dining room windows facing south are striking: the quaint waterfront houses across the street barely obscure, in the middle distance, an expanse of placid ocean that fronts Revere Beach, while off in the distance, the skyline of the city stands unobstructed, including the old Custom House Tower, the financial district centered on State Street, and even, if the telescope is pointing just so, the golden dome of the State House, where George used to work.

Felicity describes their lives as quite normal, and they are normal, or at least "normal-looking" to anyone with the same level of education and accomplishment. Felicity goes off to work each morning at a downtown law firm, their ten-year-old daughter goes off to a private school, and George works from home, mostly on a desktop computer in the living room with the commanding view.

The room itself could be a showroom of fine contemporary furniture. I would not be surprised to see Cindy Crawford stroll through, extolling the style and comfort of the place. Muted earth colors, natural wood floors, textured fabrics, fine millwork, and coordinated window treatments contribute to the inviting, if not becalming, atmosphere. It is a fine house, not merely a nice one, a beautifully situated, tastefully appointed home, undoubtedly the envy of the odd passerby.

"Our lives are so frighteningly normal in this house," Felicity told me, "that sometimes I have to stop and wonder if it really should be this way. George rarely complains about anything, and our preoccupations are pretty typical—work, family, finances, running the house, connecting with friends."

What George could complain about, and has every right to, is that over the last ten years he has lost motor control of over 99 percent of his body. His voluntary movements are now restricted to eyeblinks, very slight lip pursings, and a barely perceptible shifting of his left knee. He cannot speak, a machine breathes for him, he cannot clear his throat, scratch an itch, reposition himself in his motorized chair, or even turn his head to face a different direction when the sun coming through the wooden parlor blinds shines directly into his eyes. He can still feel everything—his sensory nerves are intact—and it is unclear if that's a blessing or a curse. Despite all this, his brain, his thought processes, his intelligence, his creativity, and his yearnings remain undiminished. He is still George, the one-time high school and college basketball star, the local kid made good, the former state director of economic development, the devoted husband and loving father. He is a man who considers himself lucky, the oddity of which dissipates, along with the oddity of Felicity's remark about the normality of their lives, when you spend some time in their home. George Kalomiris is a very lucky guy, and his home life is indeed frighteningly normal, as long as you can come to terms with the frightening part. George and Felicity have, as has their young daughter, and so have I.

George first came to me ten years ago when I was the head of neurology at St. Elizabeth's Hospital in Boston. He had just completed construction on a new house, the baby had just arrived, his career was taking off, his wife had landed a great job. Everything was coming together for him. My job was going to be to tell him how it was all going to come apart, and to walk him through the stages. Except that it did not entirely come apart. George is one of the few ALS patients I have treated who made the decision to press on no matter what. It is an agonizing choice either way, or so I thought until I met George.

It had been quite a while since I had last checked in with him. My secretary had arranged the visit for the first Saturday after New Year's Day. My ICU service was done, there was no pressing business at the hospital, so it was a good time to follow up. The day had begun bizarrely enough, with Elliott being upstaged by Sheldon's stand-up routine, followed by a testy session with my Parkinson's patient, and then a calming drive up Route 1A in light traffic. The tide was high as I drove across the narrow causeway that leads to the sheltered hamlet of Nahant, next to the city of Lynn. The deep hue of the ocean met the crisp blue of a cloudless sky in a line so sharp that I kept sneaking glimpses to my left as I drove.

When I got to George's front porch and rang the bell, I could not help glancing behind me, out over the water at the sheltered beach and the city skyline beyond it. To the left of the front door, through a bay window, I noticed that the parlor had been set up with a hospital bed and a lift. This was George's bedroom. This was his view.

When I first met him back at St. Elizabeth's, George Kalomiris looked as if he had stepped out of the pages of *Sports Illustrated*. He was six feet four inches tall and built like a tight end. His face conveyed his senses of humor, accomplishment, compassion, determination, and also fear. Today, although still recognizable, albeit with much less hair, George has been transformed, and his ability to communicate through a look, a glance, a tilt of his head is almost gone. He now sits in a mechanical chair, his head held up by a strap across his brow, his lips somewhat flaccid.

We rarely stop to think about how much of our persona is created by the forty-three or so facial muscles at our disposal, especially those that encircle our eyes. When we think of eyes, other than their color, we think mainly about their frame: the lids, lashes, and brows; a squint, a glint, an arched brow, a purposeful asymmetry. We speak with our eyes. We read other people's faces through a myriad of micro-expressions. One of the cruelties of ALS is that it not only forces its victims onto ventilators, thus robbing them of speech, but it eventually neutralizes most of the facial muscles, reducing the expressive palette to a few basic gestures.

It was hard at first to read much in George's demeanor, though I could tell that he was happy to see me. His lips barely formed the outline of a smile, his eyes evinced the stirrings of joy, his personality still came through.

Next to his wheelchair, in a compact box with tubes that run up to, into, and down through his trachea, George's ventilator emitted a rhythmic *whoosh*, not as sinister as Darth Vader's, yet still disembodied.

We took our seats in the sunroom off the kitchen—Felicity and I across the table from each other, George wheeled to the head of the table by Robert, his home health aide. Robert is one of three Ugandan assistants who trade off attending to George twenty-four hours a day, periodically suctioning out his tracheostomy tube, shifting his limbs, setting up his feedings. Robert seems perfectly suited to the task, the very picture of placidity and competence.

"We've been very lucky," Felicity said. "When George first got his trach, we hired a gentleman from Uganda to live with us and help me care for George, while I also managed a very young child and a demanding full-time job. He was here six-and-a-half days a week. His name was Elijah, and through him we got tapped into this incredible community of caring, hardworking, Ugandan health-care workers."

Felicity showed none of the care and worry I would expect, or that I had seen in spouses of other ALS sufferers. She is a handsome woman, with a face that is full, intelligent, and very pleasant. She has been through a lot, having survived her own health issues (as she would eventually tell me) over the last five years, but she wears none of this on her face and in her manner. For her, that's all in the past.

"By the time George got the tracheostomy he was worn out. In hindsight, maybe he shouldn't have waited. With the wheelchair, it took him falling and getting a concussion before he finally used it. With the feeding tube it took a ghastly amount of weight loss, and then almost choking to death before getting it. With the ventilator, it took a collapsed lung, pneumonia, and an insanely low vital capacity. He was reluctant with each new decision he faced along the way. With no available effective treatment or cure for ALS, he understood that once he sat in that wheelchair, he would never walk again; once he got that feeding tube, he would never taste food or drink again; and of course, once on the ventilator, he would never speak or breathe on his own again. He was really sick for about four months after the trach was put in, and he had a series of hospitalizations for various infections, the last being in May of '06, when he wound up in the hospital for forty-eight hours with bronchitis. But that was the last time he was in the hospital."

"Does he get simple things, like flu shots?"

"We see the pulmonologist every six months for a trach change and a respiratory assessment. And we get his pneumonia and flu shots through the Visiting Nurse Association."

"How often do they come?"

"We have an aide from VNA who comes seven days a week, and a nurse who comes about twice per month. Two people spend about three hours each morning getting him up, bathed, and dressed. It then takes an hour or two to get him back into bed at night. He definitely does best being cared for here at home, away from the hospital."

The conversation started up where it had left off years before. On its face, this was a social call. But it was just as much a follow-up visit. I came partly because I needed to know how this works, how they make it work, how others could make it work. As I expected, it works very slowly.

"Do you have a plexiglass letter board like we used to use?" I asked.

"No. We should get one. But the computer has been fabulous. He uses it to write letters, and he can e-mail. But it's a lot of work."

"Are you controlling the computer with your finger or with your head?"

"With his knee," she replied. "It's kind of weird, but he has continued to have almost imperceptible movement in one knee, and in his head."

"No finger movement?"

"*No*," George signaled with blinks.

"Yes" and "no" are a simple matter for George. In order to speak in full sentences, he uses a spelling system that divides the alphabet into five lines. He blinks to confirm the line, and his respondent then recites an alphabetical string until George blinks again.

"How do you know when to stop?" I say. "Do you go through the lines of the letter board, George, and then you stop her at one of the lines, and then she goes across the letters?"

"*Yes.*"

"George is now texting like a teenager," Felicity said. "He's gotten a little bit shorter with his messages. He used to spell out whole words and long sentences, and I would say to him, 'You don't need to spell it all out, or use prepositions.' But it's hard to break good grammar habits."

"So he's bringing you to the line? . . ."

"Yes. And he'll blink when I get to the right line. Then I go across to the letter."

"And you have an equanimity and a patience with it? You are his voice?"

"Yes."

"And this is the main thing you use with your child?"

"Yes. She's very adept, incredibly so. She stores it all in her head and rarely writes down the letters he is spelling out. My daughter read five of the seven Harry Potter books in the second grade. They're five-hundred-page novels, on average. She's an amazing reader, and I tell people it's because at the age of three she would come home from nursery school, and George would type: 'Hello, Sophie, how was your day?' Then he would hit SPEAK, so she would see what he was typing, then hear it. When she was five she learned the entire line system as a surprise for his fiftieth birthday."

It was time to get down to the business I came for.

"We had conversations when George was getting worse about what to do," I began. "There was the nodal conversation. You may be aware that while it's not exceptional to decide to keep living with ALS, it's not the normal decision, either. Most people throw in the towel. A lot of them are quite a bit older."

"We've known many ALS patients in the last ten years."

"I can imagine."

"We've met many through this journey that haven't chosen our path."

"I was acutely aware of the fact that George was young, in his thirties, and that you had a young child, you had a life to live. He had certain notions of what was important and what wasn't, what he wanted to do for himself, and what he wanted to do for his family. I'm sure you had those conversations at great length. So I'm just curious, what was that like?"

"What was it like to have those conversations?" Felicity looked at George, who rolled his eyes. "I can say a few things I remember about that. As soon as it was clear that he had ALS, George started doing the research, looking around the corner, looking the bull in the eyes, and didn't back away from trying to understand what that meant, and what was coming for him. And very early on, long before he was even in a wheelchair, he made a very clear and informed decision that he wanted to have the trach and go on a ventilator. I can recall one conversation in particular, perhaps the first on this subject. We were out on a drive, and he told me, 'I want to go on a ventilator when the time comes.' I responded that he would have my full support in whatever decision he made, and that I was glad to know that now, because we both knew there would be a lot of planning. My understanding at the time from talking to you and reading about ALS myself, was that George would lose his physical body, but his mind and senses would remain intact. I remember telling George that I fell in love with him because of his soul, and that I knew that his soul, his essence, his spirit—what defined him as a man—was going to remain intact whatever happened to him physically . . ."

She turned to face George. "Do you remember this conversation? We were on our way to Salem Hospital to see our first neurologist."

"1-A-B . . . 1-A-B-C-D-E . . . 3-K-L-M-N . . . 4-P-Q-R-S-T . . . 3-K-L-M-N-O . . . 3-K-L-M-N . . . *yes*." *Benton.*

"So it was after we had seen you, and we had a pretty good inkling of what was going on, and we were on our way back to see Dr. Benton. We were in the car talking about what it means to get to a point where you are no longer physically able to move, and you are locked into your body. Is all hope lost, and are you no longer worth anything? Are you no longer a human being? Are you not really alive at that point? George's philosophy, then and now, is that hope is never lost, although the things he hopes for may have changed with time. He tells me often that life is a daily gift, always worth living, notwithstanding the physical limitations of ALS."

When Felicity and George first visited my office together, they told me the story of the pull tabs. George was unable to grasp the baby's diaper pull tabs between his finger and his thumb, and he had said to Felicity, "Something's wrong here."

"That was October 2002, the diaper thing," Felicity reminded me. "I told him, 'Nice try, now go in there and change the diaper.' But over the winter there were other things. It would become increasingly troubling, but not enough to make us think something was wrong. I would ask, 'Did you pinch a nerve?' because George was incredibly athletic. And I remember we went to an event in January 2003, a black-tie event, and George couldn't get his cuff links on. His thumbs were weak. And at that point I finally said, 'You really should go see somebody.'"

"U . . . 3-K . . . 2-F-G-H-I . . . 1-A-B-C . . . 3-K." *You kicked?* "3-K-L-M . . ." *Me.*

"Oh, yeah, we were roughhousing on the couch with Sophie, and I kicked George in the thumb by mistake, and the thumb went right out. It popped out, because, though we did not know it at the time, all of the muscle was gone."

With seemingly infinite patience, Felicity picked up on George's subtle signals when he had something to say, and led him through the letter system. Most of his interjections were brief, but at times, I had to repress my natural urge to butt in as she painstakingly spelled out one of George's longer thoughts, a process which involved anticipation, a lot of word-filling guesses, and a good memory. As she ticked through the alphabetical system, all the time keeping her eyes on George's eyes and lips, her right hand scrawled letter sequences on a lined pad of paper that gradually filled with wayward, clipped words and sentences, cross outs, and doodles. She did all of this without a hint of frustration.

". . . O-n-e t-o t-w-o m-o-n-t-h-s? *Yes.* W-e-n-t t-o h-a-n-d s-u-r-g-e-o-n. *Right. One or two months went by . . .* a-n-d i-t n-e-v-e-r . . . h-e-a-l-e-d. *It never healed.* He's just saying that it never healed. He ended up seeing an ortho guy who was a friend of his, who had done his knees, and he was the one who observed the bilateral loss of muscle tissue, and sent him to see the neurologist right away. At the time, George was still an avid basketball player. We had just moved into this house. We moved in the summer of 2002, had the baby two weeks later, and then a year later, unfortunately, we met you. It was a fast-paced series of events that moved us from a state of incredible joy to one of profound disbelief and sadness."

Back then George had a powerful athletic body. Scattered around the living room were pictures of him from that era. The muscles were now gone.

"Where did you meet George?"

"George and I met in the late '80s. I was working in the same office building as he was. It was love at first sight, and I stalked him for a while. But then I went off to law school, George got engaged, and I got seriously involved with someone else. But we stayed in touch, our friendship deepened, and eventually we both came to realize we were meant to be together. We've known each other for a long time."

"George, do you think that religious or cultural background played a role in the decision to stay here at home on the vent?"

"L-o-v-e f-o-r l-i-f-e . . . a-n-d . . . f-a-m-i-l-y."

"If there has to be a yes or no answer to that question, I would have to say no," Felicity added.

"Did having an athletic background give you an edge, George?"

"M-a-r-t-i-a-l a-r-t-s m-o-r-e s-o . . ."

"George was a brown belt—judo. At the time the symptoms started he was doing aikido."

"Did the martial arts training help with your decision to go on the vent, or did it make it easier to manage the vent?"

"T-h-e w-i-l-l t-o c-a-r-r-y o-n . . . w-i-t-h p-u-r-p-o-s-e."

"He's saying he had reason and purpose to stick around."

"In retrospect," I asked, "did you have a realistic expectation of what it would be like to be here today?"

"It's hard to say. We had no expectations because it was so unknown. It would be as if someone had said to us: 'Felicity and George, we want you to go live on the surface of the moon for five years,' and after we lived there for five years, they said, 'Was it what you had expected?' I would have to respond, 'I wouldn't have known what to expect.' "

"George, could you have envisioned it, or do you have Felicity's perspective that it was going to be what it was going to be?"

"W-e w-e-r-e g-o-i-n-g t-o . . . a-d-a-p-t . . . n-o m-a-t-t-e-r w-h-a-t."

"There was very little anguish about what it would be," Felicity added. "In my estimation, it takes a very special person, one with great conviction and patience, to be able to say: 'This is what I want. If it turns out to be a disaster, it's going to be a disaster, and that's it.' "

"It takes a lot of courage," I added.

"It takes incredible courage. George is the bravest human being I will ever meet, but I don't think George has ever patted himself on the back about this. There are people around us who've never been able to come to terms with this: friends and family members who love him very much have had an incredibly difficult time mustering up the effort to tackle his mode of communication. And many found the tubes, the vent, the total paralysis, the seemingly complete incapacitation of a vital human being, too difficult to witness, and they eventually faded away. But those friends that were able to look into the ugly face of ALS and not turn away came to realize that the essence of George had survived this calamity, and for that they have been blessed with the ongoing gift of his love, his humor, his friendship, and an inspiration for life that comes from being around him. I know

it sounds irrational, but for George and me, and maybe in part because of our daughter, we have endeavored to live our lives normally.
I tell people all the time, we are a pretty normal couple. George is an
amazing father and an incredibly supportive husband and friend.
There's just a few things in our lives that are a little different, like,
you know, like that he is ninety-nine percent paralyzed and totally
dependent on others."

"I assume that financially it's a burden, but that there are also resources available to you."

"Yes, it has been a financial burden. But we were also very blessed.
We were at a time in our lives, just before he got sick, when we were
preparing to have a child, and were taking on a significant mortgage,
so we were planning for the future, financially speaking. He had health
insurance . . ."

"And disability insurance?"

"And long-term care. And it was really very fortuitous that we had
pursued all of that. It's very difficult financially to live like this, and
we are incredibly lucky that I have a good job, we're lucky that we
have been able to keep this house. You can't leave George by himself;
he requires twenty-four-seven care. It's a big issue for people facing
these situations to come to terms with the economic impact."

"If I hear you right," I said, "you would have made that first leap
anyway, because that was based on character, and then figured it out
later. But you've been lucky. You told me all these things early on,
and I thought, if anybody could do this, it would be you. I don't recall, was I encouraging and supportive, or did I try to discourage
you?"

"No, you did not try to discourage us. Do you remember these
conversations, George?"

"V-e-r-y w-e-l-l. R-e-c-o-m-m-e-n-d-e-d v-e-n-t. T-a-l-k-e-d a-b-o-u-t
l-i-v-e-i-n a-i-d-e . . . t-a-l-k t-o V-N-A . . . g-e-t a-n a-i-d-e."

"I don't remember you saying anything that was negative. To the
contrary, my recollection is that you clearly and nonjudgmentally dis-

cussed the choices we would face, and some implications of our decisions."

"Have you met other ALS patients through the fund-raising?"

"Yes, and this gets to your question of living with purpose. George is very busy, and has been for the last ten years. He gets up early and is engaged every day. When you no longer have a reason to get out of bed, that's when you're going to take a long look at the worth of your life. We got involved with an organization, with a group called the Angel Fund, and through that organization we've met other people afflicted with ALS, or otherwise affected by the disease one way or another."

"U . . . t-o-l-d . . . a-b-o-u-t . . . a-n-g-e-l." George interjected. "Oh!" Felicity interpreted, "You told us about the Angel Fund."

"Yes, I told you about that."

"That's been a way for us to raise money and awareness about the disease. George and I also sit on the board of an organization called Prize4Life. It's an organization that seeks to increase discovery in the field of ALS by funding prizes for people doing innovative and impactful research. We've met people through these organizations, but we're not tapped into a specific support network. We haven't had time for that. You had a big impact on George. One thing you said that has always stuck with both of us, when you were pretty sure this was ALS, was 'Go home and live. Don't go home and die.' That may be, from your perspective, just a normal thing you would say as a physician, but from our perspective, being given this kind of death sentence, it was very important advice. We took it very literally and it has saved us both in many respects. It's sad that not every physician thinks like you. We found that there's a whole generation of physicians coming up today that don't share your philosophy. They are being taught that you have to balance economics and all these other factors with the practice of medicine, which clouds their ability to effectively focus on what the patient wants and needs."

"It comes back to a rudimentary aspect of medicine," I said, "which is that the physician is the best person to advocate for *that* patient at *that* time and for *their* needs. So somebody wants to die and refuses to be like this? My role is to be their doctor. Somebody wants to keep plugging along in every way possible? My role is to help them plug along. And yet you and George are not judgmental about other people who decide to throw in the towel?"

"No. On this issue of making choices, George's view is that there is no right or wrong road. There is only the road you choose. And although many may have gone down that road before, it will always seem uncharted and untrodden to the newly diagnosed person. Some choose hospice, some do not. Some elect to have a tracheostomy and go on a ventilator, many do not. Some decide to seek experimental and unproven treatments, and many do not. George feels strongly that these choices shouldn't be judged or criticized. We are occasionally asked to talk with families that are considering vents. We try to be as helpful and informative as we can, but we understand the gravity of these decisions and avoid interjecting our personal views. George also participates in chat rooms with other patients, but . . ."

". . . n-o-t a-s o-f-t-e-n . . . a-s I u-s-e-d t-o."

"George, do you have complete control over the computer independently?" I asked.

"M-o-r-e o-r l-e-s-s."

"He just needs someone to turn it on. He doesn't read much online, even though he was an avid reader. He doesn't read books anymore, but he's got a pretty impressive entertainment system. I don't know what people afflicted with diseases like this would do without this technology. He can e-mail friends, he writes letters to our daughter, he can listen to music, download music, and do basically any project you can do with a computer. A few years ago, he composed two songs, and at one point he uploaded hundreds of our old photographs with help from aides, feeding photos into the scanner, digitizing and editing, one photo at a time."

"S-h-o-p . . ."

"And he likes to shop on Amazon, that's another thing. He's a shopaholic. I used to bitch at him: 'Look at this credit card bill! Why did you buy all this stuff?' But then I wised up. Now I say, 'I need this or that, can you go online today and get it for me?' It's great, because I hate to shop, and I don't have time, and he does all the shopping. The computer keeps him connected. My daughter and I leave here at quarter to seven every morning and don't come home until seven o'clock at night. So he's here by himself, with the aide and our dog, twelve, fourteen hours a day, and he'll often spend a decent amount of that time on the computer."

"In the end," I added, "you have to make your own decision about whether to continue. But what I tell other patients is that the decision is not irrevocable. It's not easy, but it's not irrevocable. You can change your mind."

"E-x-c-e-p-t i-f y-o-u d-e-c-i-d-e n-o-t t-o g-o o-n a v-e-n-t."

"Very funny, George."

"I think," Felicity continued, "that it is very relevant to this issue of decision-making that when George got sick, George and I were very much in love with each other, and in a strong place in our lives. We had been married for five years, we were starting out, and we were committed to each other, to our marriage, to starting a family together. It's very personal to your circumstances. We've seen families that have fallen apart—one spouse comes down with ALS, and the other gets up and leaves, or stays but lives in a state of ambivalence. It's hard to imagine, but I try not to be judgmental. The strain of such a devastating illness on caregivers is almost unfathomable. We know families dealing with ALS who have no money, no insurance, no support. We've met quite a few ALS families, people who started on this journey before us, and people who started on this journey after us, and the decision to go on a vent is part of the bigger picture of where you are in your life, who's in your life to support you, what kind of financial resources you have, who you are."

"What is your estimate of the proportion of people who have elected to go on a vent? A fifth?"

"M-a-n-y o-n-l-i-n-e t-r-y-i-n-g t-o d-e-c-i-d-e. M-a-n-y o-n-l-i-n-e o-n v-e-n-t-s, o-r d-e-a-d, w-h-o d-e-c-i-d-e-d a-g-a-i-n-s-t."

"The one person that stands out in my mind who was ambivalent about it was your friend Arnie," I said.

"When George was first diagnosed," Felicity recalled, "Arnie had had the disease for ten years. We met him at an Angel Fund event, and he was holding a bottle of beer. He said, 'I'm a slow progressor.' And George said, 'That's my buzzword, too. I'm a slow progressor.' We had great hope after that meeting because Arnie was doing so well after a decade of the disease. But two years later, I could see that Arnie was progressing rapidly, and it was heartbreaking. Arnie subsequently had a respiratory crisis at home, went into Mass General, and then got vented."

I remembered the case all too well. Arnie's respiratory crisis was one for which he and his wife were completely unprepared, and the lack of oxygen caused brain damage. "That was a disaster," I told Felicity and George. "I was in the room. It was a very painful night."

"I-f t-h-e-r-e w-a-s . . . f-o-r-m-u-l-a . . . i-t-s t-o t-r-y t-o s-t-a-y . . . o-n-e s-t-e-p a-h-e-a-d . . . i-n . . . t-e-r-m-s o-f . . . y-o-u-r d-e-c-i-s-i-o-n."

"So, yes, we know people on vents, but my impression is that it's very uncommon."

"To what extent do you avoid living in the future?" I asked.

"I'm really curious as to what George's answer to that is. I don't know about you, George, but for the last ten years, I've lived absolutely one hundred percent in the present. I had breast cancer four years ago, and thyroid cancer three years ago, and George is dealing with this. I think that our lives have leveled out, our priorities are crystal clear. Our chief priorities are maintaining our moral and spiritual compass, our health and well-being, and that of our daughter, and maintaining my livelihood. But I don't worry a lot about what is going to happen a year from now, or twenty-four months from now, or forty-eight months from now. Will we be okay at that point, or are there things that we're supposed to be doing now to make sure that we'll be okay? I don't know. I try to live in the here and now because that's what we have. I don't worry about the future. I don't know about George . . ."

"I d-o-n-t h-a-v-e . . . p-r-o-b-l-e-m l-o-o-k-i-n-g a-h-e-a-d . . . p-l-a-n-n-i-n-g."

"I go to work every day, Sophie goes to school every day. We live as normal a life as is possible. Sometimes I say to George, 'Shouldn't we be wallowing in despair, shouldn't we be anxious and depressed, something other than we are?' I think our daughter has a lot to do with the choices we make and the positive and hopeful attitude we try to bring to this situation. George reminds me often that our role is to give Sophie balloons, not anchors. We like to believe that getting up and living every day, making our lives as normal as possible, having routines in the house, boldly staring down the dragons, and getting on with the business of making life as good as we can gives her balloons, not anchors."

"From the outside," I said, "you're aware that nobody would identify this as normal. But sitting here, it doesn't seem abnormal."

"Maybe what's abnormal is the fact that we've been able to survive."

"You're doing more than surviving. You even maintain a sense of intimacy as a couple, and many couples can't manage that. You're not living parallel lives, you're living it together."

"One hundred percent. Well, maybe ninety percent. Sometimes he could kill me. George can yell at you without saying a word, and make you feel that you've been yelled at. It's his expression . . . he'll click his teeth. If something's not happening right, he'll click his teeth hard. He still has expressions, he can look angry; he can make it clear that he's very upset. Anyone would recognize it as an angry-appearing face."

"Does he have the equanimity he had before?"

"One hundred fifty percent. I find that to be completely remarkable. It's part of the frustration of living with ALS that people can't get beyond the harsh physical effects and recognize that the person that they knew for twenty-five or thirty years is very much unaltered."

"C-a-l-l m-e c-r-a-z-y b-u-t l-i-f-e i-s g-o-o-d l-i-f-e i-s g-r-e-a-t."

"One last thing," I said. "I just had a patient, a woman who I diagnosed with ALS. She was hospitalized and had all kinds of problems. She initially got a trach, but she didn't want to persist. She didn't want the trach to begin with, and I think I may have cornered her into it. It's the flip side of what you're talking about. She died in the ICU. She chose to die, and I didn't stand in her way."

"So she had the trach taken out?"

"Yes, I took the trach out at her request, at the family's request. She was able to speak for a few minutes for the first time in a while. But just a few minutes."

"You did all that?"

"I did what she needed done," I said.

"Is that comforting to you in a way?"

"Tough. Very tough. The two extremes in neurological practice, as I see it, are the result of the incredible damage that can be done to the brain and spinal cord—dehumanizing effects. One extreme is to save life at any cost; the other is to participate in ending somebody's life in order to reduce their suffering. You'd think the two extremes can be reconciled, but they can't. You have to live with both. What are you going to do? It's not my life. It's her life. I can't give or take it away from her. Are you a little bummed out that I would be part of this?"

"No. Not at all. You didn't take her life away from her. You didn't knock her off. Isn't the point you're making that it was her choice, and you were just the implementer of her wish?"

"All I could do was make her suffer more," I replied, "because of an ideological perspective that had nothing to do with the individual patient."

"No. That doesn't trouble me at all. George and I are not fixated on a notion of: 'Live, live, live, no matter what.' We just want people to be able to make their own choices."

. . .

I left, and drove back across the causeway as dusk approached, back to my wife, my house, my dog, my ham radio pals, my life, and Sheldon.

I ran into Sheldon a week later when he came back to the Brigham for a follow-up visit for his wife, who was in bad shape. Her heart was failing. Making matters even worse, his daughter was an inpatient at the Brigham due to a stroke that led to Dejerine-Roussy syndrome, a debilitating combination of aching, burning, jabbing pain that is so life-altering that it frequently leads to depression and psychiatric problems. Sheldon could only stand and watch as his daughter suffered. I had arranged to meet him at Au Bon Pain again.

"You can't make gold out of lead," he said to me, "and I've been trying for twenty-five years. But it just doesn't work. Everybody has problems, we all have struggles in life. It should be struggles maybe at first, and the rest of our lives should be lasting delight."

"This is in relation to your daughter?"

He nodded, then shrugged. No jokes.

"Listen, Sheldon. I have two questions. One is: Why did you come over to me the other day here at the hospital? The second is: What's with the comedy for you? I am a guy who's got a lot of comedic material. I grew up in a house where everybody was telling jokes in different languages and drinking schnapps until two o'clock in the morning. That was the crowd, smoking cigarettes and cigars, my uncles giving me cigars when I was nine years old. So while I don't wear it on my sleeve, I'm always interested in people who have been in the comedy-writing business. Not the flashy stuff, but just what it's about for you. But first, why did you come over to me that day, how did you know?"

"Well, when I walked over to you, when I came in, I had just gone through a terrible session with my daughter upstairs, and you look like a nice person. So I'm a people person, and I stop to tell a joke. I don't think it's inappropriate for most people, and I don't tell pornographic stories. Anyway, I came over to you because you look like a nice person."

"But do you do that all the time?"

"No. I came over because I was so overwhelmed with emotion. I just wanted to let it go. I digressed from what was happening in my life, and I went into a new life."

"And you really wrote for Youngman?"

"Henny was a dear friend of mine, rest his soul. I ate in his home innumerable times. He paid me a dollar a year so I could say I wrote jokes for Henny."

"You know," I said, "as a kid I was a waiter in the Catskills, and I saw all those guys, and I picked up their timing."

"Where did you wait?"

"Tamarack Lodge. I was at Grossinger's for one summer. So how did you get into writing jokes?"

"Just natural. I never write anything down. I'm blessed with a photographic memory. Some of it is creative, but some is material that I hear. I may embellish, I may change and so forth. But we can't endure without humor."

"That's my feeling," I said, "but as you well know, not everybody shares that. I think life is too serious to be taken entirely too seriously."

"This is true," Sheldon replied.

10

The Curse of the Werewolf

*On the front lines in the battle against
Dr. Parkinson's disease*

She is standing on the main floor of Grand Central Station, at the base of the west stairs, with her right hand on her hip and her left hand propped on the marble balustrade, frozen in the posture of a teapot. She has missed her stop, and is trying to decide whether she can find her way back without phoning someone, without asking for help. A moot point, because she knows that for the moment, in this crowded, familiar place that is now so unfamiliar, she may not be able to speak the words forming in her head. She has to suppress the panic and start over, revisit the plan for the day.

She has ventured out because no matter how badly she feels, she never misses her dance class. On some days, the effort to get there seems Sisyphean, but she manages to move the boulder to the top twice a week without fail, as if to fail once would be the end of her.

Her name means Hope, a fact that no longer holds much significance for her. She lives in the here and now, in the day to day, specifically in the tactical planning of every single day, as though a cadre of majors and generals were deploying troops, tanks, and artillery batteries around a large-scale mock-up of the theater of operations in the bogged-down war playing out in her head. But instead of a map, she has a whiteboard in her kitchen. The board for today has a bulleted list that includes pills, meals, money, workout clothes, a timetable, directions, reminders, emergency contact numbers, and a few motivational phrases:

Happiness is a decision.

What other people think of me is not my business.

Live and try.

A young woman descends the grand staircase of the concourse from the mezzanine, in thrall to the constellations mapped out on the ceiling. She is in her twenties, no longer an undergraduate, surely from out of town, well-dressed, vibrant, happy, and lost.

"Can you tell me how to get to the Times Square shuttle?" she asks Tikvah, who at first stares, as though she's looking at the young woman through inch-thick plate glass. She knows precisely the way to the subway entrance, but the words won't come. She can only point to it with her eyes. She tries to gesture, she is not mute, she knows what she wants to say, but finds herself unable to say it. The spinning motor in her head has slowed down, and the clutch won't engage. Independent of her *knowing* how to get to the subway entrance, the mechanism that effectuates the *giving* of directions isn't working. Most people do not have to think about this, but Tikvah does.

The young woman looks around, embarrassed, unnerved, and retreats without a word, but the encounter jolts Tikvah back to her own itinerary: go left, up the corridor, then down to the turnstiles. Take the 6 local to 59th Street. Uptown. Two stops.

Tikvah lives in the East 90s, in Yorkville. She is a slight person, and has a tendency to tilt forward when she walks, bending at the upper spine so that in profile she looks vaguely like a question mark. If obliged to stand still, she will episodically push one foot ahead in a half step, place her hand on her hip, and make a few rocking movements with her upper trunk, somewhat like the rhythmic to-and-fro-ing made by religious Jews when they pray, or *daven*. Given her disdain for any religious extremism, the very idea that someone might think she is davening mortifies her. While walking, she will periodically wind one or the other arm forward and to the side, turning the palm outward 90 degrees, by itself an odd gesture, but one she can morph into a natural movement by continuing it up to her head, as if she is fixing her hair. She doesn't shuffle, but her steps are short and close to the ground, and she has a tendency to look down when the terrain changes. Her foot placement is surprisingly precise. There is a delicacy to her appearance, perhaps because of her slight frame, making her look like a ballerina walking on snow.

It takes her ten minutes to reorient herself, in much the same way that it takes her ten minutes to figure out how to make a cup of coffee at home, or ten minutes to begin to get dressed, but she gets moving again, gets back on the subway. Upon emerging from the 59th Street station after a brief ride, she recognizes the terrain, the panic has subsided, and she is buoyed by the idea of stepping into the dance studio, dumping her bags, and starting to stretch. In the company of her fellow sufferers, she can let her guard down, and for a short while at least, stand without support, reach without bracing herself, move without any self-consciousness.

"Do you want to hear more?" she asks me. We are in my office, and she has been filling me in on her daily struggle.

"Yes, I do want to hear more."

"I can't put on my own makeup. I don't go out as freely as I would like. I don't have the energy, and that affects my mood. I can get really down about it."

"And the classes help?"

"Oh, yes. I have my tai chi class and my dance class. I can do things there . . . no, *we* can do things in those two classes that we cannot do in other situations. We can move without holding onto anything, do balance poses. When I walk out of there I'm high as a kite. I always make sure I get there, even if I'm exhausted. It makes a very big difference."

A bigger difference-maker is the medication I've been prescribing to replenish the dopamine level in her brain and restore motor function back to her control. We meet periodically to find the right timing and dosages. She is haunted by hallucinations, one of the common side effects of the drugs, but also of the disease itself, and unfortunately for Tikvah, her visions are not caused mainly by the pills.

"When they started," she says, "the hallucinations, what I saw at first was Queen Elizabeth and her corgis in my fireplace. Just sitting there. I also had Dick Tracy come by. He had a yellow overcoat. It was Warren Beatty. He was coming in from the kitchen looking for someone. They don't look at all fake, I can't see through them, and they move."

"Do you talk to them?"

"No. Jimmie Walker from *Good Times* has been showing up in my kitchen, and he doesn't speak and that scares me a little. He doesn't say, 'Dy-no-mite!' He comes to visit at the end of the day when I'm alone and tired. I see him more frequently when I'm tired. But I've discovered the clues that let me know when they will come. My mood will change before they visit. Sometimes the mood swings can be very frightening, but I have figured out where they come from, so I work around it."

That last phrase sums up Parkinson's disease for the million or so Americans who live with it: they work around it, with varying degrees of success, depending on the form of the disease, its progression, and the treatments available to them. There is no remission, there is no cure. There is perseverance, there is resilience, but there is also annoyance and frustration.

Tikvah is a family friend. I have known her for over forty years. I gave her the diagnosis of Parkinson's on the day of the September 11 attacks. Although I am not technically a Parkinson's specialist, she would have become my patient anyway, unlike some of the well-heeled Parkinson's patients I now follow, who came to me mostly because of a patient I first saw twenty years ago—Michael J. Fox. The quotes on Tikvah's whiteboard are from Michael's book, *Lucky Man*. The sad irony of her situation is that Michael is indeed a lucky man in a way that Tikvah is not, in a way that makes his prognosis better than hers, and it has everything to do with Dick Tracy, Jimmie Walker, and the Queen of England.

In the early 1800s, James Parkinson established a surgery at 1 Hoxton Square, in the Hackney borough of London's East End. Although trained as a surgeon, he had the keen eye of a clinician, and when he was not seeing patients in his clinic, he looked for pathological cases outside of it. This is how he came to describe three men who frequented the neighborhood, and whose alarming physical signs suggested that they suffered from a condition that medical science had yet to classify. The first case he wrote about was

> a man sixty-two years of age, the greater part of whose life had been spent as an attendant at a magistrate's office. He had suffered from the disease about eight or ten years. All the extremities were considerably agitated, the speech was very much interrupted, and the body much bowed and shaken. He walked almost entirely on the fore part of his feet, and would have fallen every step if he had not been supported by his stick. He described the disease as having come on very gradually.

The next was

> a man of about sixty-five years of age, of a remarkable athletic frame. The agitation of the limbs, and indeed of the head and of the whole body, was too vehement to allow it to be designated as trembling. He was entirely unable to walk, the body being so bowed

and the head thrown so forward as to oblige him to go on a continued run, and to employ his stick every five or six steps to force him more into an upright posture by projecting the point of it with great force against the pavement.

The last, a gentleman of seventy-two years,

first perceived weakness in the left hand and arm, and soon after found the trembling commence. In about three years afterwards the right arm became affected in a similar manner, and soon afterwards the convulsive motions affected the whole body, and began to interrupt the speech. In about three years from that time the legs became affected. At present he is almost constantly troubled with the agitation, which he describes as generally commencing in a slight degree, and gradually increasing, until it arises to such a height as to shake the room, when, by a sudden and somewhat violent change of posture, he is almost always able to stop it.

Parkinson queried some local physicians about similar cases, and in 1817, with three more patient histories in hand, he produced a monograph entitled *An Essay on the Shaking Palsy*. So keen were his observations and so crisp was his writing that the great French neurologist, Jean-Martin Charcot, would later decide that the disease Parkinson so ably described should be named in his honor. And so it was, even though it defies simple classification.

Like James Parkinson, I indulge in on-the-street diagnoses of Parkinson's disease and the spectrum of physical symptoms that come under the umbrella term *parkinsonism*. They are not hard to spot. Some sufferers are beset by slowness and rigidity, visible not only in their bearing but in their faces. Others, outside of their medication cycle, suffer from a characteristic slow tremor, often only on one side. Those in the midst of a medication cycle trade off tremors for hyperkinesias—exaggerated movements of the arms and neck—that many have incorporated into their body language. Such is the case with perhaps the world's most famous Parkinson's patient.

Michael J. Fox was first diagnosed with the disease in 1991, at the age of thirty-one, at a high point in his career. By the time he went public about his condition five years later he was starring in one of the top-rated shows on television (*Spin City*), had a string of post–*Back to the Future* movie hits under his belt, and was one of the most popular and widely recognized celebrities in America. That he happened to show up in my office before the public knew of his condition was a fluke. At the time, he was working in New York with Woody Allen on a television movie, a remake of the 1969 film *Don't Drink the Water*. His condition was still a tightly guarded secret, one that was becoming increasingly difficult to keep hidden.

The referral came from Lenny K, my former senior resident when we were training in internal medicine in San Francisco. Lenny is the most charming and committed physician in New York City, and he has managed to situate himself in a neighborhood and in a type of practice that has attracted the TV, film, and entertainment crowd. Over the years, he has sent me a stream of young, restless, bold, and beautiful daytime soap stars with the most mundane complaints, for whom a professor of neurology in Boston had a calming, therapeutic effect. This time was different.

"Allan, baby, when are you coming to New York?" That was Lenny's way of asking me to consult on a VIP patient very quickly. "Have you heard of Michael J. Fox? He's having some kind of tremor, and I think he could use a bit more attention than he's getting down here. I'm not sure what's going on. Can you see him this week?"

Celebrity entertainers show up in almost every major medical practice. Some doctors collect them. At UCLA, I recall a whiff of testosterone in the air as a group of neurologists casually referred to the stars in their care. In the New York medical scene, on the other hand, A- and B-list actors occupy a middle stratum below financial giants and politically connected families. In the Big Apple, money talks louder than fame, but in Boston this is all a novelty. When the dean of a medical school comes to my office as a patient, he usually arrives in an old Chevy wearing a sport coat with elbow patches. Same with the local pols. Even the tycoons in this town dress down.

In the same way, Michael had none of the star-making machinery about him: no entourage, no hovering manager, no bling, and no pretension. Just a regular guy. He made his own appointment, he came into my office at St. Elizabeth's Hospital without fanfare, unslung a backpack from his shoulder, and casually tossed it on the couch.

"Call me Mike," he said, holding out a hand. There was no mistaking him—it was McFly all right.

"Doc, you gotta help me fix this," he said.

"What's the problem?" I asked, even though I had already noticed the tremor in his left hand and arm.

"You want to know what the problem is? I'll tell you what the problem is. I can't read on the can in the morning. I can't read to my kid at night. I can't hold the damn book still. Really, that's all I want. I just want to be able to hold the book still so I can read to my son."

"That's all you want?"

"No. That's the obvious problem. What I really need is . . . I want you to tell me what the deal is. No one else has. I want to know everything that comes with this, or that's going to come."

He was wearing jeans and sneakers, a logo sweatshirt, sleeves rolled up to the elbow with blue oxford cloth cuffs peeking out. For all I knew, he could have skateboarded down the street behind the hospital before being ushered in via the back way to my office. No one was supposed to know he was here.

"Tell me about the medications you've been taking."

"The Sinemet helps a lot," he said, "but it wears off pretty quickly, and it's beginning to make my hands and neck move around when I don't want them to. The Artane doesn't work at all for the tremor. Until now, my strategy has been: 'Just tell me what drugs I need to take to make this go away,' but it's not working anymore."

"The tremor is not going away?"

"No. So I need you to tell me: What is it, really? What are my options, really? What's my prognosis, really? What should I do?"

His concern was justifiable, and despite his initial plea, reading to his son was the least of his worries. No one wanted to see Marty McFly or Alex Keaton or Michael Fox lose control of his body, and he knew it. That would be a tough sell in prime time.

The irony was not lost on either of us that Michael had already acted out this scenario in *Teen Wolf*, a 1985 film in which he played a high school senior who finds himself turning into a werewolf. The condition, he discovers in Act II, runs in the family (unlike his Parkinson's). In the film, he uses his wolf skills to his advantage, but in the end he decides to succeed on his own merits. He just wants to be himself. What he was telling me at our first meeting, in essence, was that he wanted the same thing.

"Doc, you gotta find a way to keep me on television for a couple more years. If you do, I'll give you a Ferrari."

So I did, which got me in some hot water, and no Ferrari.

Twenty years after my first meeting with Michael, I find myself sitting with Tikvah in my office. She is reminding me how it all started for her.

"The first thing I remember in retrospect is my pinkie finger started to wiggle. I knew it was PD right away. I'm always looking for trouble. I was geared to things medical. I went home, took a bath, and the finger was still going. So I knew. That was about twelve years ago. I did okay for a while. I could run to the subway, I could multitask. When I got the diagnosis on 9/11, I was very frightened. I said to myself, if anything goes bad, I'm cutting my throat. But it didn't hit me because I didn't feel the symptoms. They weren't painful."

"So now, a dozen years later, what is the most troubling aspect?"

"Dependence, slowness, clumsiness, no energy, mostly those things, and my mood. My mood really changes. Also, now I'm starting to have to go places *with* somebody, not on my own."

"Is it possible to articulate the sensation of the slowness, of the rigidity, of being frozen?"

"The physical stuff and the emotional stuff are very connected for me. I can get stomachaches, and think that I didn't take my medication correctly. Then I run around, and that makes it worse. It feels embarrassing. Especially my speech. I can be in midsentence, and I cannot finish the sentence. I can't find the words, and I don't know when they're going to come. I'm worried people will think I've been drinking. So if I go somewhere in the morning, I have to calculate everything, anticipate everything because of that."

"Is it word finding only?"

"No. It's both slowness and word finding. I could lose my speech and my vocabulary at any time. Sometimes people look at me like: 'Are you crazy?'"

"And with the Sinemet?"

"I feel almost normal, except that I know I have PD."

Ever since the name of the surgeon of Hoxton Square was attached to the disease over a century ago, there have been precious few breakthroughs in the treatment of Parkinson's—two to be exact, each one a game changer. The first was the accidental discovery that the surgical destruction of a structure known as the *globus pallidus* could short-circuit the signals that caused Parkinsonian tremor and rigidity. Irving Cooper, a young neurosurgeon toiling in semi-obscurity in the Bronx, perfected the technique—called a pallidotomy—in the early 1950s. His results were so dramatic and unexpected that he was accused of faking his data, and was nearly drummed out of the profession. But by the mid-1950s, the pallidotomy became the standard of care. Over the next decade, Dr. Cooper and a few other neurosurgeons experimented with alternative lesioning techniques, notably the thalamotomy—the destruction of a pea-sized portion of the thalamus, another deep-brain structure—that also proved effective in eliminating tremors. Then one morning in the fall of 1968, after having performed over six thousand of these procedures, as many as six per day, Dr. Cooper walked into his clinic and announced that all such surgeries would stop, effective immediately. The reason was the arrival of a nonsurgical alternative: the drug called L-dopa, the secret ingredient of Sinemet.

Sinemet is in many ways a wonder drug. Its origins go back to 1967, when a Harvard-trained physician named George Cotzias found that administration of the chemical L-dopa in carefully timed doses relieved most Parkinson symptoms. L-dopa, or levodopa, crosses the blood-brain barrier and is converted into dopamine, the neurotransmitter whose curtailed production is a signature feature of Parkinson's. Over the next decade, a series of hybrid drugs that combined L-dopa with a targeted delivery agent hit the market. One of these was Sinemet. Although it had only a modest effect on tremor, it became the drug of choice for most other Parkinson's symptoms. With surgery no longer considered optimal, or even advisable, the question for neurologists boiled down to this: for any given patient, which of the L-dopa drugs should be chosen, how big should the doses be, and how frequently should they be given? To figure that out, I sometimes had to dabble in a little field research.

The Teamsters at the Chelsea Piers were not so much intimidating as annoying. It was clear that they felt entitled to scoff at anyone who was not in the "business," meaning show business, and meaning me. Even though I was expected at the television studio, it took three phone calls at the gate to finally shove my way past the muscle.

For a sophisticated and high-budget television show, the set of *Spin City* looked thrown together, rough around the edges. But edges don't show up on camera. The soundstage was on my left as I entered. A brightly lit, cavernous space, it contained only one enclosed room, a kind of trailer. Michael J. Fox was in it, half lying on a forlorn leather couch in spartan surroundings. There was a small carpet, a wooden chair off to the side, a simple desk, a small fridge, and little else. He offered me a club soda.

I had come to talk about his medication schedule. It had been two years since he had shown up in my office, and he was now starring in a new sitcom, his condition still a tightly guarded secret. I was there to see how we might organize his meds for maximum benefit while they shot his scenes.

When I walked in he was in an "off" period, his speech monotone and mumbly, his face expressionless, his movements reluctant, and his posture poor. We discussed the Sinemet not in terms of doses, but colors. He knew that if he took a full blue, he would be waving his arms like a wild man within twenty minutes, and would be of no use on the set. On the other hand, half a yellow would not give him enough fluidity or the full range of expression that he needed to play his role. That's the downside of Sinemet: the trade-off between tremor and unseemly dyskinesias. The drug provides an inverted arc of an effect: a big shot of L-dopa knocks out the symptoms, affording a twenty-minute window of fluidity and fluency. If too much is taken, the dose will overshoot the mark, producing the wild gesticulations and head twists. When the L-dopa begins to wear off, the descent is even quicker: a rapid reentry into a state of torpor marked by a frozen affect, the opposite of Michael's on-screen persona.

I thought that one yellow followed by another half-yellow, twenty to thirty minutes later, would get him into the comfort zone without shooting him into the twilight zone. He had been experimenting with something close to this, and it seemed like a good plan. I hung around to witness the effect, and it worked. He could sense that the Sinemet had kicked in when he felt a vague background aura of charged energy, and he emerged from his dressing room and ran through his scenes with no tremor and only minimal dyskinesia. I could see his actorly control over every posture and movement, how he disguised small dyskinesias behind the frenetic mannerisms that had become part of the character he was playing. I kept an eye on my watch during the shooting sequence, and at about eighteen minutes into it, I could see that he was going to crash. Michael sensed it too, and strolled off the set with a minute or two to spare, with no drama, as calmly as if it happened five times a day, which it did. After nailing his final line, he returned to the little room for a between-rounds conference with his corner man, the keeper of the yellows, blues, and other colors of the dopamine spectrum.

"What does it feel like when it wears off?" I ask Tikvah about the Sinemet.

"I feel agitated, with stomach pain. Then it knocks me over the head, and I have no energy. That's when I know the meds are wearing off. I have no choice, and I have to sit down, and an hour or two later I'll still be sitting there, wondering how I got there."

"If you had patients with Parkinson's, how would you advise them?"

"They have to rest. They need to change their self-image. The whole family does. You already know that the disease doesn't just affect individuals, it impacts the entire family."

I have known Tikvah's husband Saul for decades. I know their son and daughter, and I know how unhappy her illness makes them. It comes out when Saul tells me how sad he is for her, and how difficult it is to accommodate not just her slowness, but the daily struggle she endures. Their son can roll with it, but their daughter initially had a contact aversion to the disease. She's scared, she doesn't know what to do about her mother, what to do when she gets confused, what happens when she's really disabled, and I am inferring from my experience with other families: *Is it going to happen to me?*

"Does it help that Michael J. Fox represents the disease publicly?" I ask her.

"Oh, yes. I have a lot of respect for him. He just goes out and does what he does, and doesn't care what anybody says. That's the sense I have of him. And having him as a spokesperson, he's a regular guy, not some esoteric guy. That makes a big difference. He has no shame about it, which is great."

I hear the same sentiment from a lot of my patients for whom self-consciousness is part of the Parkinson's struggle. They all say that it helps to have Michael out in front of them. When he announced his condition to the world in 1997, he told reporters that he planned to act for three more years. I told him he was probably good for another ten. In fact, he is still going strong, three Golden Globe and eight Emmy award nominations later. But it has taken more than Sinemet.

Fox called his book *Lucky Man*, and he is lucky in several ways. Not many people could get such a diagnosis at age thirty-one, and over the ensuing two decades stay at the top of their profession, win awards, and thrive in a celebrity fishbowl. Yet he may not quite appreciate just how lucky he is in one sense: no one else with a bread-and-butter Parkinson's could have lasted as long as he has. I give him credit for his fortitude, and he deserves it. Managing that kind of medication schedule requires discipline of the highest order. But it has also required one more thing that Michael could not control: remaining medication-responsive ten years out from initiating Sinemet. That's a rarity, not unheard of, but unusual.

In 1997, I first broached the possibility of brain surgery with Michael. At that time, neurosurgeons in Europe were experimenting with an extension of the Cooper procedure in which, instead of merely making a hole in the thalamus or globus pallidus, they placed an electrode at the site, and ran wires from the electrode through the skull to a small stimulator, essentially a pacemaker, implanted under the collar bone. It was called deep brain stimulation, or DBS. At that time, the technique and the equipment were primitive by today's standards, and not yet approved by the FDA. The procedure offered little in the way of customization, and Michael was adamant that he did not want wires coming out of his head. But he was still a good candidate for the old Cooper procedure—the lesion without the implant—and this made sense given his unique needs and the time frame. He was still dealing mostly with tremor, something a thalamotomy could handle very well.

What's more, during the surgery, we could test whether the lesion would work before having to make it. Of course, once the hole was made in the brain, there was no filling it in, it would be irrevocable. Only one question remained: I needed to find out whether, if Michael had the thalamotomy now, he would later be able to get a deep brain stimulator. I talked it through with some top-flight neuroanatomists in Boston, as well as two of the major neurosurgeons in Sweden, where the field was quite advanced. What would happen in someone who had a hole in their thalamus, who later got an electrode put in the globus pallidus or subthalamus? It may sound wonky, but everything was at stake.

"No problem," they said. In the end, I presented the thalamotomy to Michael as the best option, even though at the time it had become passé among U.S. neurosurgeons.

"This is a good idea," I told him, "and it doesn't close the door on anything else. That's the consensus. I think you've got almost nothing to lose."

"Let's do it," he said.

But where? Several surgeons in Boston were doing functional neurosurgery for Parkinson's disease, but few were doing surgery for tremor. The bigger problem was getting a highly visible patient out of the public eye. I resorted to an old friend, Dr. Bruce Cook, who had helped me on some tough cases. Bruce had admitting privileges at a smaller hospital north of Boston that was affiliated with ours. The hospital had recently installed a $10 million, state-of-the-art neurosurgery suite.

The scene conjured up Dr. Frankenstein's lab more than an ultramodern operating theater, not because of the hospital, but because of the novelty of the surgery itself. It was intensely computer-aided. Using anatomical landmarks on an MRI, Bruce guided a test electrode through a tiny plastic tube down into Michael's right thalamus. He warmed it up a few degrees, and the tremor, once violent enough at times to knock a glass off the table, stopped in its tracks. That was the sign that the location was correct, that the procedure would work.

There were five of us in the room: Fox, fully conscious and responsive the entire time, Bruce Cook, a surgical assistant, a nurse, and me. Nothing dramatic occurred until the end of the fourth hour, when Bruce said, "Okay, Michael, create a tremor in your left arm for me."

"I'm trying," he said.

"Keep trying."

With evident frustration, Michael said, "I can't."

Bruce abruptly stepped back, whipped off his gloves, and said, "We're done." The tremor was gone. The next day Michael walked out with a sweatshirt over his head and went back to New York, a new man.

The criticism was predictable, and not long in coming. Journalists who can't get close to celebrity patients do the next best thing: they pigeonhole a talking head into Monday morning quarterbacking a case of which they have no firsthand knowledge.

"Some in the medical profession are questioning why celebrities who could afford the best of everything would select a small community hospital that sits twenty-five miles up the road from some of the best teaching hospitals in the country." Thus began the carping in one local paper.

"I found it a little odd," a local neurosurgeon was quoted as saying.

"My initial response?" said another, "I found it a little disconcerting." He publicly wondered "whether the thalamotomy is the best procedure for the actor, since he is so young and his disease will likely progress and include other symptoms not alleviated by that particular surgery."

My answer? Yes, it was the best procedure, and Michael agrees.

"That thalamotomy," he recently told me, "I don't think for a second that that thalamotomy was a mistake. It was fantastic. And people say to me, 'Why don't you want to have a DBS?' and I say because on my best day without it, I'm ten times better than the people who I see who have had the surgery. They may have less tremor and less dyskinesias, but they have this kind of slowness and this whisper and this fragility. So I'm fine, my thalamotomy was great. It couldn't alter progression on my right side, but it never was intended to."

Back in my office, I have to explain to Tikvah that she is not a candidate for DBS, or even a thalamotomy. No one does thalamotomies or pallidotomies anymore, but that's beside the point. She can't get a stimulator.

"One of the contraindications," I tell her, "is that you can't have any cognitive difficulty or hallucinations that are not solely attributable to Sinemet. Now, if you have hallucinations that abate when you reduce the dose, you're still a candidate. But in your case, because of the confusion and visions, you are not. There are too many bad outcomes with confused people."

Tikvah's hallucinations are partly Sinemet-related, are mainly nocturnal, and are a common feature of diseases related to Parkinson's, such as Lewy-body dementia. She has enough cognitive difficulty that she has begun to struggle with executive organization, sequencing, and planning. That's what keeps knocking her down day after day, that's why she has a whiteboard in her kitchen that details her daily routine, that's why she cannot get the surgery. There is a high likelihood that the surgery would bring on the kind of mental slowness that Michael alluded to. It's not her memory problem that is so worrisome, rather her inability to organize.

"Who are you?" I ask her. "How do you see yourself?"

She replies, "I'm a wife, I'm a mother, I'm an observer of mankind. But what I am—just like you—you're a doctor, and I'm a psychiatric social worker who's seen a lot of whitewater go under the bridge. I've seen people out on the ledge, I've seen people die, I've seen people not wanting to die. That's who I am."

Though she was forced to retire after thirty years as a therapist, she still views herself and others through the lens of her profession.

"I'm in two support groups," she reminds me, "one for PD and the other for children of Holocaust survivors. Both are about survival. The woman who brought me into the PD group was a remarkable woman, an artist, very successful, very strong, very well known in Manhattan, a force to be reckoned with. She could go to a store on the Upper East Side, and come back with a new member for the group. She didn't survive. She killed herself. I have two insights on that, one professional and one personal. It's an awful thing either way. She had a rough time of it, more so than the rest of us. Some people might say maybe they were going to kill themselves, and then not do it. But she was a determined soul, she looked it up online, she measured the drugs out, and there was suicide in her family, so it wasn't a foreign concept."

"Was she very disabled?" I ask.

"Yes. She was the most disabled person in the group, although we didn't notice that because she was so competent and so talented and so friendly."

"From your professional experience, do you think she got profoundly depressed, or was it a conscious effort to find a way out?"

"A conscious effort, yes. She was very angry, very irate at the disease, what it had done to her, how badly her family had responded. She was an artist and did really good work. All they did was criticize her. That's how she viewed it. The group as a whole got very upset with her family, which I think ruined it for everyone. Then one day she was gone." She goes silent for a moment, then resumes.

"Personally, in my bones, I've increasingly felt, partly from my parents' experiences, partly from my upbringing in Israel, that the Holocaust is very much a part of how I interact with this disease. Maybe it's a less explicit analogy for others with PD. The sense of imprisonment, the sense that you're not going to get out of it, that the disease is in charge. You can do certain things, you can get along, go out to a movie, but in the end, the disease will take over and you will be tired, and it will wear you down. The suffering, the not knowing what you will end up with. No relief. Who's going to take care of you? Who's going to want to take care of you when you are in that deformed condition? We talk about this in my support group. Thank God for the group."

Lenny K sent Michael J. Fox to me most likely because he looks at me as the last Mohican. He knew I would take a detailed history and try to tailor a treatment to Fox's needs, and stick with him. That's my value added. There are people who may think I'm the world's expert on Guillain-Barré syndrome, people who may think I'm the world's expert on concussion and head trauma, and people who may think I'm the world's expert on Parkinson's disease, mainly because I'm an expert in none of those, but I'm a good doctor for all of them. Michael didn't need the world's leading Parkinson's expert. That guy is required only for a case that no one else can figure out. Everybody else out on the street with Parkinson's simply needs a good neurologist. Michael just needed someone good enough, and he got me.

Michael starred in a new show that premiered in the fall of 2013, twenty-two years after his initial diagnosis. His foundation has raised a quarter of a billion dollars for Parkinson's research. He is no longer my patient, but he checks in from time to time.

"I took up golf in my forties, with Parkinson's," he recently told me, "so that's optimistic beyond anything I could ever do. What I've had a good time with is watching able-bodied golfers agonize over missed shots and putts, and they get so angry, and they attach shame to it, and they say, 'I'm sorry, I'm so sorry I missed that,' and I laugh and shank it into the woods, and I say, 'What's my handicap? Isn't it obvious? It's not a number, it's what I deal with.' And I feel so bad for these people who attach shame and attach failure to a game. It's their life, but I refuse to do it, I refuse to apologize for my Parkinson's. If I manipulate it or cover it up, it's not out of shame, but it's about being able to do what I do to the best of my ability, and accomplish what I need to accomplish. And I recognize what it has cost me, I know the loss."

"What do you mean? What does it cost you?"

"I can't run, I had to stop training for the marathon, I couldn't do that and other things, so I applied it to writing a book. In a large way, you were responsible for me looking at it empirically, and looking at it in the nonemotional way that I really needed to do, to separate all of the feelings of fear and all of the feelings of concern and shame and guilt, the idea that I must have done something to deserve this, and then: 'What am I putting my family through?' You helped me put all of that stuff aside, so I could ask myself, 'What is it I am dealing with? What is happening in my brain? Why is it happening? Can I stop it from happening? Can I, now that it's out there, exercise my privileges in the wider world to move forward toward a cure.' All of that came from our meeting."

Tikvah is still visited regularly by Dick Tracy and Jimmie Walker, and her form of the disease is different enough from Michael's that I can hold out only the slimmest hope of a surgical solution. After our appointment ends, she finds her way back to the reception area, and I drop by the conference room to hear a visiting speaker, Mike Okun, from the University of Florida, who has come to the Brigham to give a talk about deep brain stimulation. His concluding statement strikes a familiar chord: "After twenty years of DBS, it is time to stop thinking about which [surgical] target is best, and ask which symptom the patient needs fixed. We have to move from disease treatment to symptom treatment."

The remark takes me back to the future, back to my recommendation for Michael's surgery: treat the symptom, not the disease; treat the patient, not the disease. That's the mantra for Parkinson's patients— patch them up and send them back out there, and be waiting when they come back at the end of the next round.

That's me in the corner, still waiting for my Ferrari.

11

For the Want of a Nail

A hard-knock lesson on the way
to the morgue

I was in the kitchen preparing a snack when my cell phone rang. It was the radiology department. "Your patient, Mr. Connaway, has a spinal epidural abscess," the neuroradiologist said. "It's massive."

This confirmed my worst fear. Connaway was going to be a quadriplegic at best (the correct term is *tetraplegic*), or dead within days. And the thought lodged in my brain: *Had I killed him?* But he wasn't dead yet. Maybe there was still time to save him.

From my house it is about a twenty-minute drive to the hospital in afternoon traffic. I didn't have twenty minutes, so I sailed through two stop signs and one red light before I caught a break: an idling Boston patrol car on Commonwealth Avenue. I pulled alongside, rolled down the window and said, "Look, I'm a doctor at the Brigham. I know this is unusual, but I have to get to the hospital NOW. Like ten minutes ago." And he said, "No sweat, man. Follow me." With the siren wailing, we zipped through the evening rush hour traffic, defying red lights, down a back way I didn't even know existed. Less than ten minutes later I abandoned my car at valet parking, ran up to radiology, took a look at the MRI, then ran up to the ICU. Harry Connaway and his wife were there, just as I had left them two hours before. Harry was in terrible shape, had been ill served by a dozen practitioners, including me, and had little hope for a recovery, or even survival. But the existence of a chance, however slight, meant that I had to do everything I could for him.

Four days earlier, Connaway had been playing tennis. He was pretty good at it, and played regularly. He lived in an expensive house in an exclusive suburb. Recently retired, he was reaping the rewards of four decades of very stressful work as an investment banker. He was conscientious about his health, took a daily cholesterol drug, and flew through his annual medical exams. He was understandably baffled when a progressive numbness took hold in his legs, and then spread to his arms and hands over the course of a couple of days. His wife called 911, and Harry was brought to the emergency room at a local community hospital.

His most visible symptoms—generalized weakness of the arms and legs that, by the time he arrived at the hospital, made it impossible for him to stand, accompanied by loss of reflexes—pointed to a problem in the peripheral nervous system, most likely the result of Guillain-Barré syndrome, a treatable form of paralysis due to an inflammation of the nerves. It can cause a rapidly progressive weakness, and the disease is right in my wheelhouse: serious but manageable if caught in time. Joseph Heller had it, as did Andy Griffith. Both men recovered.

Harry Connaway's legs were weak, but not paralyzed. His bladder wasn't quite right, but it was functioning. Having settled on this diagnosis as the most probable one, his doctors were preparing to treat him when he suddenly became very unstable. His blood pressure and heart rate began to bounce between dangerously low and absurdly high. On day three, when he lost bladder function, they threw in the towel. "This is beyond us," they said. "Let's send him to a bigger place."

I got the call from Harry's physician early that evening, and said, "Sure, send him over. It sounds like it could be Guillain-Barré. We'll treat him tomorrow morning." I saw no reason to do it in the middle of the night. It wasn't justified. A few hours delay in starting the IV drip would not make a difference.

So Harry Connaway was wheeled through our doors at 11:45 p.m. on the third day of his crisis. Stanley, the first-year resident who was handling admissions, called me at home to say that he had just examined a sixty-eight-year-old man who had lost his reflexes. He said nothing about sensory levels, a standard neurological exam finding that tests for numbness along the spine. A specific transition point from normal sensation to numbness is a characteristic indicator for spinal cord problems. Stanley did mention a fever and high white count, but I downplayed those facts in my mind.

"Keep him supported," I said. "We'll see him in the morning."

I can't blame Stanley. He got almost every detail right. The problem was my rigid thinking. I should have asked for more detail, but I was stuck on a snap diagnosis, and because I did not ask, the rest of the night passed with no action taken.

Harry Connaway's fortunes started spiraling downward upon his arrival at his local hospital, where the radiologist missed something on his MRI. In his defense, it wasn't something obvious. Harry had a rare condition. Still, for want of the correct reading, three days were lost. Upon his arrival at the Brigham, that mistake was compounded by me, by my failure to ask the right questions immediately, leading to more errors of omission. In Harry's case, I latched onto an explanation that happens to be a specialty of mine: Guillain-Barré syndrome. As it turned out, my patient would have been better served by a fox than by a hedgehog.

The interdependency that is built into a modern hospital, our reliance on checklists, hierarchies, consultations, and each other, protects us individually from taking all of the blame when things go wrong. We can and do make mistakes, but in theory someone is there to catch and correct them. In practice, the system of checks and balances can fail at every level, and a low probability event, a succession of unlikely failures, can be set into motion. In the case of Harry Connaway, a cascade of errors of omission began to fall like dominoes.

I saw Connaway the morning after his admission to the Brigham at around 6:30 a.m. I was serving as the attending in the neuro-ICU. My first impression, based on the man and his chart, was that Harry was the equivalent of me in another profession: hardworking, committed, active, with a wide circle of friends. At some point the similarities ended. Even in his weakened state, crammed into a hospital bed, I could tell that Harry Connaway was a tall, strapping man, the kind who projects physical vitality. His face had the confident and commanding aspect of a hawk. I still didn't know about his sensory level. Because we were making speed rounds, we weren't yet collecting the kind of detailed data that we normally do on the ICU. Too many things were going on at once. Our paramount goal was to check on all of the patients in the unit just to make sure they would live through the next few hours. When the patient count rises, as it had on that morning, rounding can get disorganized. It would take us three hours to get back to Harry for a more thorough exam. We were assuming his condition was urgent, but not critical. I was still working with a diagnosis of Guillain-Barré.

Stanley, who had admitted Harry, was the only one of us who had examined him carefully. He was the member of the team who knew him best. At 9:30 that morning he had to sign out because he had reached his mandated eighty-hour limit for the week. That was unfortunate. The one person who embodied most of the detail in Harry's case was leaving. At the end of speed rounds, he gave the team his handoff and went home. Distilled in that handoff were Connaway's records from the community hospital. These included an MRI of his lower back, recorded as normal and dutifully passed on as such to us. I have no idea why they had ordered a lumbar MRI at all. A scan

can occasionally be helpful in confirming a diagnosis of Guillain-Barré, but it is not a standard test. When Joelle, the senior ICU resident, looked up the scan in the patient notes, she repeated to me that the scan was normal. I didn't know whether she had seen the film itself or was merely reading the summary, and I didn't ask. Although she was nominally in charge, I said, "Let me examine the guy, I know a lot about Guillain-Barré, and I'm always interested to see variations." Already I was misleading myself.

By the dictates of the scientific method, we are obliged to seek evidence to disprove our assumptions rather than verify them. If you have a theory, you are not supposed to fixate only on the evidence supporting it. Admittedly, you will rarely see Sherlock do this on *Masterpiece Mystery!* The Holmesian thrill lies in instantly seizing on the right explanation from the barest facts. What Holmes should be doing, and what Dr. Watson usually suggests, is exploring other possible explanations. Instead of assuming that he's always right, Watson might say, "Just this once, consider the possibility that you might be wrong." Without realizing it, I, too, had thrown the scientific method out the window.

I noted that Harry could barely move his face, and assumed it meant that his symptoms couldn't just be a problem at the neck or in the spinal cord. Also, his limb tone was gone. He was floppy and had no reflexes, both of which are highly characteristic of Guillain-Barré. Third, he did not have Babinski signs, a toe reflex which is probably the most typical feature of compression of the spinal cord. So I said, "Okay, that all fits together. It's GBS. Let's get the electrical nerve conduction tests and see how bad it is." I told Joelle that I wanted it done that afternoon. An alternate theory, the one that I was overlooking, was that some of the examination findings could have been caused by the medications Connaway had been given. They could easily have wiped out his facial tone and muscle tone completely.

At midday, a team showed up to run the electrical nerve tests. The technologist called me an hour later and said, "It wasn't easy to do because he was in the unit, and there was a lot of electrical noise and people coming and going, but the bottom line is he has no excitable nerves. When I tried to stimulate a nerve in his limb, I didn't get any response. He probably has really bad Guillain-Barré because the nerves are shot."

It all seemed to fit. The technologist was merely echoing my own thoughts: a case of *axonal* Guillain-Barré, in which the nerve fibers (or axons) are wiped out by an immune reaction. The senior EMG physician called me with the same conclusion.

Deductive reasoning, as opposed to inductive reasoning, is what detectives and diagnosticians should be doing. Deduction works from general facts toward specific conclusions. If the "facts" really are facts, the conclusions have to be true. But induction—a quicker and much more practical method of reasoning that everyone uses every day—can lead to errors. That's because an inductive process infers a conclusion, but doesn't prove it. If the nerve conduction tests are normal, for example, the patient cannot have Guillain-Barré. But the inverse of that statement—that an abnormal test (like Harry's) *must* be Guillain-Barré—does not logically follow. My initial error was excessive dependence on induction due to time pressures, as well as hubris. In practice, neither Sherlock Holmes nor I have the luxury of time to indulge in the true deductive method, and when the clock is ticking we have to rely on intuition. But sometimes intuition can let you down.

When Harry's wife showed up early that afternoon, I finally got a chance to hear a firsthand account of the onset of the paralysis. The next fifteen minutes represented the height of clinical skill: a mental juggling act that involved listening with an informed ear while classifying everything relevant, filtering out everything irrelevant, and creating a hypothesis to test against the examination findings. The time had come to apply the synthesizing, value-added process that would bring me back to the scientific method. The challenge was that my source of information was an increasingly and justifiably frantic spouse who was picking up on the urgency of the situation through my facial expressions.

"Well, you know," she said, "he was doing alright, then he had some trouble walking, then he felt feverish and just terrible, and he couldn't urinate. Then he suddenly got a very bad pain between his shoulder blades, and it got worse and worse over the day. Then the pain went away."

The acute pain, while not typical, could possibly fit with Guillain-Barré. I know because I had written the paper on it. It's called the *coup de poignard*, or stroke of the dagger. But the fever and bladder issue didn't fit. That was highly suspicious. At this point it occurred to me that something wasn't right.

"Let's start over," I said to Joelle.

"Why?"

"Something's off here. In fact, let's look at those films." It was now 3:00 in the afternoon. The other residents on the team demurred, saying, "But they were normal."

"Have you seen them?" I asked.

"No."

"Well then let's have a look, because if the nerve roots are enhancing, that's got to be GBS, and we could maybe put this to bed."

On Harry's MRI everything looked completely normal, if not pristine. The nerve roots were enhancing a bit. But at the very top edge of the scan, I noticed a small white ball, a mere eighth of an inch in diameter, which seemed to be pressing on the bottom of the spinal cord. (Most people don't understand that the spinal cord doesn't go all the way down the spinal column, but stops in the upper back.)

I said, "What's that?"

Joelle replied, "It's just an artifact," by which she meant a by-product of the imaging process, like a dead bug in the lens.

"I don't think so. That looks real. I think it's an epidural abscess."

Only now, at 3:30, did it occur to us that Harry Connaway's problem might be far more dire than we originally thought. I immediately took the MRI down to one of our radiologists, and said, "I want you to find your chief and reread this. I'm not going to tell you what's there. Just reread it and call me."

For the moment, there was nothing more to do. The radiologist at the local hospital had written up a report that failed to raise a red flag, and until thirty minutes ago we had relied on that report. Until now, no one at the Brigham had looked closely at the scan. I still thought there was a good chance our radiologist might say, "Oh, yeah, it's just a blip," but that was hope staving off conviction. I went home to grab a snack and change my shirt. If they found nothing, then my day was over. Instead, when the phone rang at 4:15 p.m., I found out it was just starting.

"Your patient, Mr. Connaway, has a spinal epidural abscess. It's massive."

"Oh, shit!"

The bad news, I told Harry and his wife after my police escort, was that he had an aggressive infection within his spinal column that had begun to put pressure on the spinal cord, cutting off its blood supply. His spinal cord was slowly being strangled. This explained all of his symptoms.

It took me almost eighteen hours to catch a mistake that should have been caught when he was admitted, or even earlier when the ambulance delivered him to his local hospital. Had the technician there moved the image frame even a centimeter higher, no one could have missed it: the lower limit of an enormous abscess running from his neck down to his mid-back. It was all pus, and it appeared as a mere blur in the upper edge of the frame where no one was expecting to see anything. It was missed by a radiologist, it was missed by my junior resident, who possibly didn't even see it, it was missed by an intern, who did look at it on the previous evening. If someone had told me there was an abnormality, I would have said, "Drop everything. Let's get moving here!" Instead, the whole thing was now a horrifying mess. Every neurologist should keep epidural abscesses in mind for this very reason. They're not common; in fact they are quite rare, but they can be devastating if not caught immediately.

I had to get Connaway to an MRI machine at all costs. We needed to be sure about the abscess and find its extent so that it could be drained by a surgeon. His blood pressure had been fluctuating wildly all day, his fever was going up and down, he had coded twice, he had four tubes infusing fluids and blood-pressure-supporting medication into his arms because he was that sick, almost too sick to be shifted off the bed and onto a stretcher to bring to the machine seven floors down. Harry was on the brink. I said to the techs, "I don't care. It might kill him, but if we miss this we miss our only shot." Joelle placed the order. They called for him to come down at 5:30, which would have been okay, but the nurse said, "He's crashing, we can't bring him down now. He'll have another cardiac arrest." So they bumped him to the end of the list.

Two hours gone since the order was placed, then three hours, and they still couldn't get him to the scanner. He was too unstable. Four hours, and he was still crashing. I kept insisting that we had to make this happen now. I finally said, "I'll go down with him. We have to get this scan because I need to tell a surgeon what he has, what its extent is, and where he needs to operate. He might die down there, but we have no other choice."

So three of us—the nurse, Joelle, and I—brought him down with a cardiac monitor and a bag that we compressed by hand to support his breathing. We did a quick-and-dirty MRI, and saw the whole extent of the abscess: It was about as bad as it gets.

Once we had Harry back in his room and were joined by his wife, I couldn't dwell on what-ifs. There was still much to be done. "I think, unfortunately, that the infection has been around long enough, including today, that it is at risk of damaging his spinal cord irrevocably. So we need to drain the pus. The abscess needs to come out or he'll be paralyzed, and wither away, and die. I can't promise you that he's not going to be paralyzed, even with an operation."

It was 7:30 that evening when I called around to my neurosurgical colleagues. At the Brigham, the responsibility for taking care of spine problems alternates between the neurosurgeons and the orthopedists, both of whom are extremely capable. I called my go-to guy first, a neurosurgeon. I was fairly sure he wouldn't be anxious to come in and perform a ten-hour procedure overnight. He said, "I'd love to do it, but it's their turn," meaning the orthopedists. I then got hold of a senior resident in orthopedics, who tried to talk me out of doing it late at night because, as he said, "What's the difference between now and the morning?" To which I replied, "It's now or never. I'm going to transfer him to another hospital if you can't do it." Facing that threat, the resident called his attending physician, who is a very good guy, and he came in that evening at 9:30, looked at the film, started the case an hour later, and finished it at 8:00 the next morning. I was waiting for him to come out of the OR.

"He's still paralyzed."

Three hours later, back at the bedside, I had another talk with Mrs. Connaway, who was understandably shaken, and said that her husband absolutely did not want to be kept alive like this. "Five days ago he was playing tennis," she said. "If there's a prospect that he's going to spend the rest of his life a quadriplegic, we need to let him go."

Then it was Harry's turn. He was so sick that it took all of my skill to get a little bit of yes or no out of him by lipreading. The tube in his throat and his by-now complete paralysis made it almost impossible for him to communicate. I had to get very close to him to be sure I had his full attention and that I could interpret his responses, a proximity that would have been inappropriate in any other setting. I leaned steeply over the bed, and brought our faces perpendicular to each other. I could tell he was in pain. His lips were moving, his shoulders barely shrugged. I could sense his animation and agitation. The abscess had penetrated all the way up to the nerves that innervate the neck muscles, so far up that he could only communicate by blinking and barely moving his lips. I spent forty minutes with him, finally establishing that if he was going to be a tetraplegic, he just wanted it over with. I managed to persuade him to wait a day or two.

On the following day, not only was there no improvement, but when we did another MRI scan it was clear that the spinal cord had been turned to mush, that it had been completely deprived of blood flow because of the infection around it, and he was, metaphorically, cooked. Moreover, he was cooked at a level that's almost incomprehensible: at the junction of the brain stem coming out of the skull, where it connects to the spinal cord. It doesn't get any crueler. If it went any higher into the cranium, it would crash the medulla and kill him. That's probably what was happening to him intermittently as his blood pressure careened from one extreme to the other. The team met and we concurred: Harry Connaway would be irrevocably and completely paralyzed. I would tell his wife, then talk to him in order to get another affirmation that he didn't want anything more done to keep him alive. I did all of the talking. He responded as best he could, well enough to make it clear that he did not want to continue.

Because one of the results of an acute transection of a spinal cord that high up is an inability to sustain blood pressure, the decision was not about removing him from life support. All we had to do was stop raising the intravenous pressors, the medication that was keeping his blood pressure up. We wouldn't withdraw anything. When his blood pressure started to decline, we wouldn't raise it as we had before. Harry understood that. So we watched as his BP declined more and more. Two hours later, he died.

I told Harry's wife everything, pretty much. I didn't say that we blew it in quite those brutal terms, but I said he had a very extensive abscess, an aggressive infection, that it had grown and grown despite the surgery, despite the antibiotics, and that his spinal cord was thoroughly destroyed. I told her that she had made the right decision by encouraging us to let him die peacefully, because he would have lived as a quadriplegic on a ventilator, if he had survived at all. He made

clear in his determination that he did not want that under any circumstances, and therefore it was justified to back off and quietly let
him die. I told her that. I did not elaborate on the timing that I was
still beating myself up about. I did tell her that I wished we could
have done better for him, and she was extraordinarily gracious. But I
didn't do what we are told to do, openly acknowledge our mistakes,
because although I know now what the mistakes were, I couldn't then
be sure they had altered the outcome. It may well have been too late
for the surgery to have made a difference. Nor would a nonspecialist
understand the subtlety of how this had played out.

"It would help us very much," I told her, "to be able to do an autopsy. It could provide closure in a case like this." She agreed.

As residents learn the practice of medicine in a hospital setting, the
last skill they acquire is in many ways the most important one, and
the most difficult. They need to master the executive skills required
to make happen the things that need to happen for their patients. If
you need the country's best hand surgeon to come in and reattach a
hand at two o'clock in the morning, you'd better be able to make that
happen. If your patient needs a tumor removed from an optic nerve
before she goes blind, you've got to be able to make the calls and pull
the right strings. And if you need to take your patient to another
hospital to make it happen, then so be it. You'll catch hell for it, but
you've got to save your guy.

Of course, I didn't save my guy. Maybe I could have. But I couldn't
be sure until we had the results of the autopsy. It would have been a
lot easier to bury the whole thing and leave a big question mark. But
I could not allow that to happen, not only for myself, but because I
owed it to Harry. In my mind, I might have made an error, but I
could still make something positive come of it. So the next day I took
Joelle, Stanley, a junior resident, and a medical student down to the
morgue where they could see firsthand what happens when you make
a mistake. Hannah, who was on neuropathology rotation, also tagged

along. They all needed to see, as did I, whether we could have saved poor Connaway.

Part of my job is to impress upon the residents and the medical students that they are part of a tradition, and that the history of their profession is not merely of anecdotal interest. Unlike the study of, say, mathematics or physics, where it matters little to the high-level practitioner how Newton "discovered" the calculus or his theory of gravity, the practice of medicine benefits from revisiting the discoveries of the past. For a physician, seeing further means looking *over* the shoulders of giants.

The history of medical discovery is instructive, and no one person embodies that process more than William Osler, a Canadian physician born in 1849, who is considered by many to be the father of modern medicine. It was Osler who created the first medical residency program at Johns Hopkins, and established the foundations of medical training that are still largely in use today. As a result of his example, showing up at an autopsy is Oslerian, using the postmortem as a method of improving your own skills is Oslerian, teaching from that experience is Oslerian.

It may have been Osler who set the standards for deep medical training, but for me, the giant who brought clinical neuropathology to its most refined stage, not just looking at slides under a microscope and deciding whether a tumor is good or bad, but understanding how disease affects the nervous system, was Raymond Adams, who trained me at Mass General. Adams fully understood that he was working in the Oslerian tradition in that he was using pathology as the basic science of neurology, and neuropathology was to be his main vehicle for teaching. He was right, and is still right. To be a truly advanced neurologist, one could argue, you have to know about the sophisticated genetics of neurological disease and the cell biology of neurological disease and the immunology of neurological disease. That is the case for the full-fledged professional. But there is a sweet spot for the clinician, and you don't have to go quite that far to hit it. That sweet spot is still this kind of autopsy: the Osler-inspired neuropathology that Raymond Adams perfected. A lot of people might argue with that, but it's true. Anyone who trained under Adams spent a full year doing neuropathology. This doesn't happen anymore, but as a resident, I removed almost two hundred brains at autopsies in a single year. We took out the brains and the spinal cords, put them in formalin, waited two weeks, took out the material, diced it up, gave the technicians the slides we wanted to be prepared, and told them what stains were needed. It was a core part of the training.

These days, I occasionally remove a brain for the residents in order to point things out to them when there isn't a lot at stake. But they don't get the chance to do it themselves. Very few autopsies are performed today for a variety of reasons, mostly due to a misguided faith in the power of scans, and because Medicare and Medicaid do not pay for them. The hospital has to swallow the cost, and it is a very expensive proposition. And physicians, because they no longer tend to think of the autopsy as the immediate extension of clinical work, generally do not think about asking for permission to perform one. Where it was once a routine thing to do, it is now perceived as ghoulish.

In Osler's day, a faculty member would not get his hands dirty at an autopsy. The cutting would have been done either by a resident or, in most cases, by an assistant called a *diener,* the German word for servant, or more precisely, a corpse servant. Today, the job is no longer menial; the diener is a trained specialist. From a medico-legal point of view, I wasn't supposed to interfere in her work, and I didn't. I asked for a full spinal removal. I needed to see the spinal column, leaving as much of the bone intact as possible. When she laid it out on the table, we could see the large column of pus. Its uppermost extent was clearly visible, as was its direct juxtaposition to the spinal cord. The intensity of the infection had blocked all the blood vessels. That in turn had caused the spinal cord to turn to cream pudding.

It is impossible to say what Harry's spinal cord looked like when he had arrived at the Brigham, but from the autopsy, it looked like it had been infarcted days earlier. The evolution of the infection was evident, but its cause would remain a mystery. In most instances of epidural abscess, the origin is an infection elsewhere in the body. Unless you introduce bacteria during an operation directly, it can only reach the spinal column via the bloodstream from somewhere else. The diener and the pathologists looked at every organ, every piece of skin, every orifice, even the anal fissures. There were no sores, no pneumonia, nothing in the heart, no endocarditis of the heart valves. That was odd.

Another anomaly was that in most cases the pus is concentrated in one area, so that the abscess is fairly restricted, usually to a few segments of the spinal column. In Harry's case there was no clear point of origin. The infection was uniform along the entire length of the spinal canal, from the base of the brain all the way down to the tailbone. The epidural space was bursting at the seams. There were pockets of pus from stem to stern, pockets between every rib. The diener didn't have to lance them. The cord itself had been completely ravaged. Microbiologists have a healthy respect for bacterial infections because they move very quickly and do a lot of tissue damage, yet they are treatable. But it was clear that despite forty-eight hours of intensive antibiotics and a surgical draining, this was a very aggressive *staphylococcus aureus*, and we probably couldn't have done much about it. That was the one finding that assuaged my guilt.

William Osler built his reputation on correlating what he had seen during the patient's life with what he saw at the autopsy. He used to go down in his frock coat in midsummer into these hot basement rooms with his acolytes, and stand there while the organs were removed, and he would handle them, inspect them, and discuss what he had seen in his examination of the live patient, and how it related to what they now saw; what they had missed and, equally important, what they had gotten right. He would try to create a feedback loop that completed each patient's story. That feedback loop has now been curtailed by imaging, the new pathobiology that, by default, has replaced the autopsy with CT scans, MRIs, and ultrasounds. Before imaging and after Osler, it was considered the ultimate badge of a quality clinician to learn directly from the autopsy, which was the only way to look at the anatomical structures at that time. It meant you were willing to confront your mistakes. That was the classical approach.

Today, that approach has been compromised and compressed. In its place we have a Morbidity and Mortality conference every fourth Friday, in which the senior resident who has run the service, either in the ward or in the ICU, presents a synopsis of that month's cases. By dictum of the Massachusetts Department of Public Health, these M&Ms are top-heavy on statistics: How many admissions? How many discharges? How many deaths? How many unexpected complications? How many patients who were discharged had to be transferred out or transferred back? Medicare and other agencies want to know these metrics, even though they have no pedagogical or practical value. After that, the senior resident presents her biggest cases and complications. Only then, and in less than five minutes, would a case like Harry Connaway's be discussed. It is not a great opportunity for reflection.

I take some solace in the fact that four residents and a medical student got to experience the Oslerian perspective. That exercise, as much as any, is what will turn them into doctors instead of automatons. First, there's a very intangible phenomenon of having seen a patient hours before when he was alive, and then seeing him dead, with the cause of death flayed out in front of you, knowing with precision how your involvement contributed to this result. Second, you will eternally cycle back to this possibility; it will be embedded in your memory in such a way that you will never miss it again. That was Osler's innovation. And third, these students and residents didn't have to read anything about epidural abscess because it was all there in plain sight: how the abscess affects the spinal cord, how it spreads up and down the spinal column, what its potential sources are, what effect it has on the neurological exam. An abstraction of that experience ends up in a textbook, but is no substitute for seeing it. Everyone had studied it in medical school, it might even have shown up on board exams, but none of the residents had experienced it. They couldn't have, because this case was one in a million, so extreme that it simulated another illness.

My sole consolation is that they will never through their careers miss another abscess. Nor again, I hope, will I.

12

The Eyes Have It

When is somebody not dead yet?

In Boston, during the interminable leafless months, there is a big difference between a day that feels cold and one that only looks cold. Given a choice, I'll take the former. The latter just makes you shrivel. Driving to the hospital down Route 9, I could see obscenely large icicles suppurating from ice dams on eaves and dormers, frozen mist glaciating tree branches and power lines, and coagulating slush clotting sidewalks and sewers. My son, who was doing a surgical internship in Florida, had been beefing about the rainstorms down there, so I stopped the car along the Riverway and took a picture with my phone to reassure him he had the better deal. Down there, the rain quickly disappears down the storm drains, but way up north, in the slough of the Fenway, the ice which had overtaken the city was here to stay. By tomorrow's inbound commute, it would produce my next patient.

It was a bad omen when my beeper went off the following morning at 6:10 as I warmed up my car in the garage. Trey, my senior ICU fellow, asked me to meet him in the emergency room to save time on an admission. "A guy fell on Comm Ave this morning and cracked his head," he informed me when I rang him back. "He's got a big subdural, but the neurosurgeons don't want him because he may be too far gone."

Comm Ave is Commonwealth Avenue, and it snakes through Newton, past Boston College, then through Boston University, on down to the Public Garden. A *subdural* is short for a subdural hematoma, a brain hemorrhage typical of traumatic brain injuries, caused in this case when the skull hit the sidewalk and the brain caromed off the inside of the skull, in the process tearing veins that run across its surface.

When I arrived at the Emergency Department, I found the poor guy—a very thin, elfin-looking man, with pallid skin and short, whitish, sparse hair—breathing on a ventilator. According to his driver's license, his name was Mike Kavanagh, and he was an organ donor. A huge gash decorated his scalp, with dry-crusted blood cascading onto the bedsheets like a frozen brackish waterfall. His neck was wrapped in a leather Miami J collar that made him look like a gaunt, yoked horse.

"He was intubated in the field with a GCS of five," Trey dourly pronounced, as if that would capture everything about the case. It rarely does. The Glasgow Coma Scale was devised as an assessment of consciousness, especially for the use of first responders in cases of head trauma. It runs from a high value of 15 for a normal person, down to 3, indicating deep coma. "The EMTs said he slipped on the ice," Trey continued. "Someone saw it and told them that."

"Let's see what we have," I replied.

When I pulled back the sheets I saw Mike's thin frame: short, maybe five feet four inches, small-boned, tautly muscled, almost anorexic. He could have been a jockey. His skin was blanch white, with a fair number of age spots but also a farrago of freckles on his cheeks and upper back. His hair, even whiter, accented rather than covered his gouged scalp. Definitely Irish. If his name wasn't a giveaway, the Sinn Féin tattoos on his deltoids were. No outward sign of life other than the rising and falling of his chest with the ventilator breaths.

A few days earlier, my colleague, Martin Samuels, the head of the neurology department, had given a lunchtime lecture on consciousness, coma, and death by brain criteria. Here, I thought, was the poster boy for the third part of that talk. Marty prepares all of his talks, no matter the audience, as though he were addressing the annual meeting of the American Academy of Neurology, leavening insight with humor and technical minutia with lively anecdotes. Trey had been at the talk, and even though I had heard it before, I had tagged along, not just for the free lunch, but because I enjoy the way Marty launches into the topic:

> *Imagine two scenarios. In one, the viewer, perhaps yourself, sees a dog being hit by a car, being thrown thirty feet to the side of the road, where it lies motionless, probably dead. In the other, you watch a fly land on a countertop, where it is swatted and effectively crushed. Most definitely dead. Why do we agonize over the dog but not the fly? The philosopher says, "Because the dog is conscious and the fly is not." How do we know? Because the dog has eyes that look at you. The secret to consciousness is in the eyes. If the creature has humanoid eyes and it looks at you, then it is conscious. And if it doesn't, it isn't. The compound eyes of the fly do not convey any feeling to us of consciousness. Same with the worm. Philosophers and neurologists have come to the same conclusion with regard to consciousness, and that is that the secret to consciousness is in the eyes. If you understand the eyes, you will understand a lot about consciousness. In fact if you know a lot about the eyes—in great detail—then you're practically a neurologist. This is not chance.*

"Dr. Ropper, are you going to put gloves on?" The nurses always have to ask me this because I am obdurately inclined to forget. Gloves are a modern and quite sensible innovation, but I have always had trouble finding size sevens for my smallish hands. The cheaper size S

gloves that come in boxes are too small, and the Ms make me fumbly.
I grabbed a pair of Ms to set a good example to the junior residents.
No need to worry about fumbling with this patient. The moment I
pulled Mike's lids up to check the reaction of his pupils to my flashlight,
I knew that he was dead. No tone, no resistance in his eyelids, well
beyond the sleeplike state of coma.

> *Remember, an overdose of barbiturates will make you look abso-*
> *lutely dead—no pupils, no eye movement, no vestibular ocular*
> *reflex, no calorics, no EEG, nothing—until you crawl out of the*
> *grave. So until you know with reasonable certainty that there are*
> *no drugs, that there has not been any severe hypothermia, and*
> *the criteria are fulfilled, don't pronounce anyone dead.*

"Has he gotten paralytic drugs or sedation from the ED guys?" I
did not want to be fooled by an artificial death created by the medications that doctors use to facilitate intubation.

"Nope." Trey was rhythmically shifting his considerable weight
from side to side, making me nervous. Trey is at least six feet six
inches, a recognizable cell tower even when spotted from a great distance. We had twenty-two patients to round on upstairs in the neuro-
ICU, and at the moment there were no beds for the thin man. Not
only would Trey have to clean up all the medical problems from the
overnight shift, but one of us (him) would be stuck making calls all
morning, putting the squeeze on our colleagues upstairs to send one
of their patients from the ICU to the ward in order to free up a bed.
Shifting your weight is something you do when the worst part of your
day is ahead of you rather than behind you. I tried to recall whom I
had done a favor for in the past week, who might oblige me with a
move. I also wondered if there was any point to it, whether Mike was
already gone and we could simply house him on the ward.

"Look, Trey, maybe he's brain dead, and we can dispense with the
ICU bed."

"Yeah, but the ward team won't take him on a ventilator."

Mike's pupils were enlarged, without a hint of constriction. I went right up against his face, the ventilator tubing digging into my neck, looking for even a glimmer of movement as I swung the flashlight in and out of his eye.

"Nothing on the other side, either."

Don't let yourself be forced to pronounce death by brain criteria until you're ready, no matter what they try to do. The neurology service is put in the middle here, in the Emergency Department and the Intensive Care Units, for various reasons, some of which have to do with organ transplantation, some of them with family pressure, some with the anxiety of caregivers (that is, other doctors).

By the time we could move a post-op patient to free up a room for Irish Mike Kavanagh, it was easily 10 a.m., and I was mad at myself for not doing a better exam in the ED to see if our guy was truly brain dead. That diagnosis is transformative. The same warm body from ten seconds ago undergoes a state change, much like the transition of a liquid into a solid, and once pronounced officially dead, is entitled to all the rights and privileges thereof, which is to say, none whatsoever. When you are dead you cease to be a person, and you become an object. You no longer have possessions, a future, even a present, only a past. But Mike was not dead yet, not legally, not biologically, not until we said he was. He was still warm, still breathing (with assistance), still digesting his last meal.

With a bed now ready, we swung the gurney onto the service elevator, trailed by a respiratory therapist and an ED nurse, both pushing rolling IV poles, and looking like two subway straphangers on the ride upstairs. Whenever I feel I haven't wrangled myself into enough exercise in the previous few days, I look for opportunities to join in physical tasks around the unit. Transferring an unresponsive patient from the gurney to the bed is just such a ritualistic dance: turn the body away from the bed up on its side, slip a varnished board with a cutout handle on one end under him (*Why only one handle?* I always think), roll him back onto the board, grab the sheets from the opposite side of the bed, wait for someone to count "one-two-three," and pull the sheet across the board and onto the bed with the body.

As we shifted him, I perceived an all-too-familiar feeling of "dead weight" that gave further credence to my initial impression of death.

The cranial nerves are key because they come out of the brain stem segmentally, and the big question is: is the brain stem involved? That's why we begin with the eyes. The eyes are the secret to unconsciousness. We can't test smell in these patients because that requires cooperation. Remember that smelling salts do not test smell; they test pain, the fifth cranial nerve. You can test everything else, but do you? You don't have to. The eyes include cranial nerves 2, 3, 4, and 6, so you can go from 2 to 6 just by looking at the eyes. Whether the person gags or swallows when you move the tube depends on the level of anesthetic and many other functions.

"Watch the vent, dammit!" Trey shouted at the nurses. "It's pulling the endotracheal tube out."

What I noticed, and what I'm sure the others occupied in the dance did not, was the complete lack of a cough, a grunt, or even a quickening of breath coming from the patient. Sliding a tube up and down the trachea is one of the most noxious stimuli that can be applied to any live body, so noxious that it stimulates certain obligatory reactions. Awake but paralyzed persons tear profusely. Comatose people flinch and cough weakly. The dead are dead to it, and our guy hadn't even flinched.

Once we had him situated in the bed, it was time for a curious piece of neurology: the brain death exam. This involves a sequence of tests designed to confirm a suspicion that the brain is not working on any level. There are five features that confirm a diagnosis of death by brain criteria, a few surrogate features that accomplish the same task, and a few that exclude the diagnosis. If you get a perfect score on the brain-death examination—five yeses—you can officially pronounce the patient as dead as Jacob Marley: as dead as a doornail. There is no other neurologic situation in which you would push a body to the extremes that these tests require, in particular, shutting off the ventilator to see whether someone can breathe on their own. It could kill someone who wasn't already dead.

We don't call it "brain death." We call it "death by brain criteria." To use the term "brain dead" confuses the public because they ask the question: "If it's only brain death, what is alive? Are the kidneys alive?" I don't use the term alive *with regard to the kidneys or the skin. The person is in the brain, and virtually everybody in every culture agrees with that. It's death, just like death by cardiac criteria. Our job is to make sure that people don't abuse it.*

In the United States and most European countries, the brain-death exam has become a generally agreed-upon series of bedside clinical tests that start at the top of brain, the cerebral hemispheres, and sequentially test the function of each part below: the midbrain, the pons, and finally, the medulla. While a dead brain is one thing, and a fairly easy thing to confirm, a dead person is another thing, and Marty was indulging in a bit of oversimplification when he said that "the person is in the brain, and virtually everybody in every culture agrees with that." It would be more accurate to say that the medical establishment, via the medical schools, agrees with that. But the question hangs out there like a Cadillac on cinderblocks: What does it mean to be brain dead? Is that really the same as being just plain old dead?

Here was Mike Kavanagh, a warm body, chest rising and falling rhythmically, oxygenated blood coursing through his arteries, finger-nails growing, facial hair still sprouting, digestive tract still sending nutrients into his bloodstream, all of his vital organs save one most definitely alive. But both Trey and I were sure that he was in fact dead, or, more to the point, that after due process we would be sign-ing a certificate that established not just the "fact" of his death, but the precise moment of it. The inherent absurdity in such cases is that whatever had happened to Mike out on Comm Ave, and whatever was happening inside his body and inside his skull right at that moment, his death and the instant of its occurrence were up to us. According to Catholic doctrine, Trey and I had the power and the duty to decide when to release the soul of Michael Joseph James Kavanagh unto his Maker.

"Go ahead, Trey, and I'll watch and record."

But before we could begin, Elliott walked in, always to be depended on to get us coffee whether we want it or not, sometimes appearing out of nowhere with a nugget of information, whether we want it or not.

"He's a pervert you know," he said.

"What the hell are you talking about?" I replied.

"Nancy, the swing nurse, told me. He lives in her neighborhood in Malden, and everyone went ballistic when he moved in. He's a child molester."

"Jesus Christ!"

Trey gave me a long, probing look. I pinched the bridge of my nose hard between thumb and forefinger. As if on cue, Elliott grabbed his coffee, puffed out his cheeks in self-exoneration, and, with a baleful grin said, "Well . . . enjoy!" And he left.

Trey turned to me and said, "Okay, so is he dead *now?*"

Knowing when someone's alive and knowing when someone's dead: it's one of the most important jobs that doctors do. If we can't do that, we can't do anything.

Marty's words kept ringing in my head. It is one of our most important functions, and in most cases, it is unambiguous: death by cardiac criteria is a well-established standard that applies to the vast majority of patients who die in a hospital, or en route to a hospital, or at home. Death by brain criteria is another story.

When my department chairman, Raymond Adams, agreed to serve as the only neurologically experienced physician on a committee headed by Dr. Henry K. Beecher from the Harvard Medical School in 1968, he expected little controversy. The group was convened by the dean of the Medical School to look into establishing ways of determining whether a coma could be deemed irreversible, thus the title of the paper they produced: "A Definition of Irreversible Coma." The dean also anticipated little criticism, but there was a storm of outrage from the Catholic Church and from many physicians and philosophers because of the paper's first sentence: "Our primary purpose is to define irreversible coma as a new criterion of death." It was instantly recognized that there could be a conflict of interest from the emerging field of organ transplantation, which had originated at the Brigham in 1954, when Nobel Prize winner Joseph Murray performed the first successful kidney transplant.

It was partly in response to the advent of organ transplantation that the Vatican began looking into the issue of brain death in the early 1960s. The Beecher Committee report caught the College of Cardinals off guard, but it became the touchstone for all that has happened since. In 1981, a presidential commission codified the Beecher Committee's finding as the Uniform Determination of Death Act, establishing the justification for organ retrieval as death by "whole brain criteria," or "the irreversible cessation of all functions of the entire brain." The imprimatur of Harvard Medical School and a presidential commission did not resolve the question, as it turned out. The Beecher Committee concluded that "medical opinion is ready to accept new criteria for pronouncing death to have occurred in an individual sustaining irreversible coma as a result of permanent brain damage."

The Presidential Commission said, "We are going to define death of the whole brain as the death of the organism." But as at least one critic would note, "brain death is a conclusion in search of a justification."

Medicine on the whole has no trouble making the determination of brain death. That is a technical matter. The nagging issue is whether a warm, pink, pulsating, live-looking body can or should be called dead. All of the organs are viable. The body could go through the onset of puberty, it could gestate an infant. There are such cases on record. What the Beecher Committee accomplished was to find a good reason not to utilize resources on people who would unquestionably die without ever regaining consciousness. Being able to change their classification and call them dead had virtue for society. They said, in effect, "It's not living if your brain is irrevocably gone; it's not living, so you can go ahead and take the organs."

They had a clear mandate to protect the physician. They recommended, for example, that the patient be declared dead before the respirator is disconnected, so as to avoid the appearance of pulling the plug on a living person. They also recommended that any physicians involved in transplanting the organs recuse themselves from the decision process. But they were guided by practical motives, not strictly scientific ones, and the legacy of the Beecher Committee and the Presidential Commission have now trapped us. What somebody needed to say was: we're going to have a societal shift, and if your brain is so irrevocably and totally damaged that there is no hope of recovery, and it's total (so that there won't be any quibbling), then the patient is in a state where it is reasonable to do organ transplants. Calling it death was the problem.

In a moment of great clarity, the Catholic Church signed on to the idea that brain death is death. In 2006, I served on the panel of the Pontifical Academy of Sciences at the Vatican that produced a monograph entitled: "The Signs of Death." We discussed every angel one could fit on the head of a pin, including whether "death" meant that every cell in the body must be dead, an obvious extreme given that hair and nails continue to grow after death. The monograph began with a quote from St. Augustine:

> Thus, when the functions of the brain, which are, so to speak, at the service of the soul, cease completely because of some defect or perturbation—since the messengers of the sensations and the agents of movement no longer act—it is as if the soul was no longer present, and was not in the body, and it has gone away.

In the end, the church adopted the view that brain death is death, and reaffirmed Augustine's basic and ancient view, while reserving a place for the soul separate from the body. And yet within the Church there remain strong opponents of this brain-death perspective, even of removing organs from the brain dead, and these forces are again beckoning at the pope's door.

I once had a patient, a member of Hell's Angels, who was shot in the face while driving his Harley down the interstate at eighty miles per hour—with a shotgun, no less—who then went off the road and creamed his entire cranium (no helmet), and was quite obviously brain dead. Some of his brain matter was left on the road. They might as well have decapitated him, except that when he was placed on a ventilator—intubated—his heart still pumped and the body was kept alive. Here was this outlaw, a tough guy, maybe a sociopath, and ironically, it said "organ donor" on his driver's license. In the end, he turned out to be a humanitarian.

There was no question of transplanting his corneas. They were no longer there, and that was the problem: How do you know a person like that *is* brain dead without the all-disclosing eyes? We couldn't examine his pupils. We couldn't examine his eye movements or his corneal reflexes. I recall thinking, *Oh my God, you don't want to get a diagnosis of brain death wrong because it would be like committing an innocent man to the electric chair. How am I going to finesse this?*

We could show that he had apnea (that he couldn't breathe on his own), but that's just one of the five elements of the tests for death by brain criteria. The Presidential Commission said that in certain circumstances you can use surrogate tests. So we did an EEG. He had barely enough scalp on which to place the leads, and the readout was flat, or, in tech-speak, isoelectric. We did a cerebral blood flow scan. The examination took hours, not because his face was blown off, but because I had a tough time persuading the nuclear medicine people to come in and do the test at night for a dead guy. They said, "We'll come in and do it in the morning."

I said, "They don't want to lose his organs. Come in now."

They said, "Oh, c'mon!"

"No, get your ass in here and do it or I'll keep kicking it up the chain until somebody does." Finally their man came in, injected technetium, a radioactive element which circulates to show if there's any blood going into the brain. There wasn't.

No one agonized over the case, possibly because the biker no longer had eyes, unlike Marty's example of the dog hit by the car. There wasn't much left of him as a recognizable person. But inside, he possessed pristine organs that did an inconceivable amount of good, perhaps even a redemptive amount of good.

Could we say the same for Mike Kavanagh?

"By the way . . ." I said to the head ICU nurse.

"Yes, I know," she interrupted, coldly finishing my thought, "he's a donor."

The news about Mike's past had spread instantly, and a pall had settled over the unit. Once someone is pegged as brain dead, the collective investment of psychic energy in the presence of the body deflates, the motivating principle resets, and that patient becomes marginalized on the ward. A rescue mission becomes a salvage mission: we're just preserving organs. But around Mike the child molester, after the initial shock wore off, the effect became palpable. Staff people walked around the entrance to his room in an arc, as if there were cold air coming out of it. His very presence was an insult. Everyone wanted it to be somebody else's problem, but it was ours.

"How about that, Trey? The guy was an angel."

Trey harrumphed, reached under the covers, and pinched the skin on Mike's abdomen.

"Jesus, Trey! A bit coarse, don't you think?"

"Sorry, but it's better than twisting his nipples." He was referring to an obsolete practice from my generation of neurologists, now roundly considered utterly barbaric. The pinch elicited no movement, not even a brief jerk of the torso or limbs. Next came a more conventional and ostensibly more humane stimulus of applying serious pressure on a knuckle of each limb using the shaft of a reflex hammer, the neurologist's favorite weapon. Women neurologists, I have noticed, tend to press harder than men, as if to insure that no one is getting out alive. In this instance there was not a whit of movement. All but a fully paralyzed, comatose patient would exhibit a straightening of the arms and pushing backward as the shoulders rotate internally. But here: nothing, no cerebral response.

Trey then peered in at the pupils. "Round, eight millimeters," he shouted.

"Did you measure or are you guessing?" Up went Trey's eyebrows and out came a round laminated pupil gauge with a series of black circles of increasing size. Trey held it up to the patient's eye.

"Okay, *seven* millimeters and not reactive." The midbrain could now be checked off the list, given that it controls pupillary size and reaction to light.

"Ambiguous," I said. If the pupils are too small or too big, they indicate a remnant of brain function in the pons. Nine millimeters would have been unusual in a true case of brain death. Seven was okay, but not conclusive.

"Do you want me to do calorics or doll's eyes?" Trey asked, referring to two methods of making the eyes move from one side to the other, thereby testing the integrity of the pons, the middle part of the brain stem. Because the patient had a collar on, and might have a broken neck, we couldn't move his head from side to side, so we did the caloric reflex: squirting ice water into one ear, then the other. This provides a potent stimulus to eye movements through a hard-wired circuit in the brainstem, if it's functioning at all.

"*Nada.*"

Now the moment of truth, the apnea test. Will he breathe? "How do you like to do it, Dr. Ropper?"

"Preoxygenate him."

The person must have an apnea test. Then you can prove to your-self that the whole brain, including the brainstem, is gone. Just remember, when you take a patient off a ventilator, either for an apnea test or after a declaration of death, make sure that family members are out of sight, and forewarn the nurses. Many brain dead patients, once the ventilator is removed, exhibit the so-called Lazarus sign, where their arms spontaneously contract and their hands come up to their chest as though they're grasp-ing for the endotracheal tube. It's creepy no matter how many times you've seen it.

This test is the big one. It grew out of Moses Maimonides' practice of holding up a glass to see if the breath fogs it. The object is to see whether the patient will breathe on his own. We sent 100 percent oxygen through Mike Kavanagh's lungs for two minutes, enough to sustain his heart and blood pressure without a ventilator for the next ten minutes, then shut off the ventilator.

Silence. I could hear my pocket watch ticking. As Trey and I watched closely, we could see a few arching movements in Mike's back, definitely not Lazarus signs, but something not entirely compatible with brain death. We waited. With the palm of my hand acting as Maimonides' mirror, I felt air moving almost imperceptibly in and out of the ventilator tube. Was he breathing? Trey was convinced that he wasn't, but he was already convinced that the guy was dead. It was important to be sure.

It can be useful to have a nemesis, ideally an arch-nemesis. Mine is Shewmon. Although not really my nemesis, he is a real person with a point of contention that has pitted us against each other for more than a decade.

Alan Shewmon, a respected neurologist and professor at UCLA Medical Center, has the audacity to claim that brain death is not death. He didn't always believe this, but his worldview changed about a decade ago when he was presented with a patient, a fourteen-year-old boy, who suffered a severe head trauma after jumping onto the hood of a slow-moving car, falling off, hitting his head on the curb, and eventually being confirmed as brain dead. Yet he "lived" for another sixty-three days on a respirator and vital fluids. Dr. Shewmon was called in to examine the boy, and he agreed with the brain death determination. By the standards of the State of California, the boy was a corpse, but at the family's insistence, the body was kept going in a facility under the care of nurses who were baffled and unsettled by that diagnosis. Shewmon reported that the boy began to go through puberty and eventually died, not from his head injury, but of pneumonia.

How long can a body survive in a brain-dead state? Not indefinitely, but longer than most people think. That is Shewmon's central point: How do we classify such a body, because death by brain criteria is not the same as death of the biological organism? It might well be considered death by psychological, sociological, legal, or religious standards, but as Shewmon notes, if you brought a biologist into the nursing facility to see the fourteen-year-old boy, and, telling him or her nothing else, simply asked whether this was a living organism or a dead organism, any biologist would have to say that the boy was alive—severely compromised, but living.

Shewmon's point is a subtle one, and although it does not diverge very far from true orthodoxy, from the orthodoxy Trey defended unflinchingly, it diverges enough to make many neurologists uncomfortable. If we accept Shewmon's point with no qualification, it would mean that transplant surgeons are killing people, murdering them, and that we, the neurologists who sign off on it by declaring people brain dead, are accessories.

Shewmon himself is not making that argument, nor am I. Nor is Shewmon really my arch-nemesis. I once staked out an opposing view on this one point, but so did Shewmon himself. I was well aware that the moment we declared Mike Kavanagh brain dead, a transplant surgeon would walk through the door and cut a lymph node out of Mike's neck without anesthesia. After some quick tests, Mike would be whisked off to a specialized operating room, his muscles would be paralyzed, and, again without anesthesia, his kidneys, liver, heart, corneas, skin, and bone marrow would be removed and carefully preserved. He would not spend very much time in that awkward state of being: a warm, pulsating body that is legally a corpse. The removal of vital organs would quickly render him dead by any standard, yet during that short interval between the ICU and the OR, Mike Kavanagh would exist in a kind of limbo: legally dead but not biologically dead. He would be a "neomort," to use a term invented by the psychiatrist William Gaylin in 1974, three years before Robin Cook's novel *Coma* became a movie, and turned brain death and neomorts into an unsettling film-noir scenario.

It can be humbling to cede a point to an opponent, unless you come to realize that he is not an opponent after all, rather that he is a scientist doing what scientists are supposed to do: question the status quo. Alan Shewmon claims, rightly, that there are three concepts of death: a biological one that speaks of the organism, a psychological one that speaks of the person, and a sociological one that speaks of the legal person. Currently, the tests for death by brain criteria and their rationale are entirely biological. But Shewmon argues that equating brain death with the death of the person crosses the line into another domain. If we are going to pronounce someone dead because their brain has ceased functioning, we have to make that argument on psychological, sociological, or other grounds, and not on biological ones. His point about the biologist, while dramatic, is something of a canard. A good biologist would see through it as a trap. If you have a body that is a living organism, but is the equivalent of a worm, it is not necessarily a living human being.

Ten minutes were up, and the result was conclusive. Mike Kavanagh had failed the apnea test.

"That's a wrap," Trey said, snapping off his gloves.

"Is it? If he's dead, in what sense is he dead?"

"In the dead sense," Trey replied.

"Well, his brain may be dead, but his other organs are alive. They can be transplanted."

"But they're just organs. Organs can be sustained, even grown outside of a body, independent of a body."

"The gash on his neck where the transplant surgeon cuts out a lymph node would heal."

"Those are just cells," Trey countered. "They're on automatic pilot. You provide them with blood, they keep going, but there's nothing meaningful going on."

"But if Mike Kavanagh were a pregnant woman," I said, "we could keep him alive in order to bring the baby to term. What could be more meaningful than that? My point is that we just engaged in an operational decision, not a biological one. The end result is still correct, but we shouldn't pat ourselves on the back and say that we have come to an ontological certainty. We need to be honest about what we're doing. His brain may be dead, but the rest of him is not dead, and we can use the rest of him. I have no problem with what we're doing, but I agree with Shewmon: we should think it through more carefully."

Trey paused, and said, "And that's what we just did, right?"

"You're not buying it, are you?"

"No," he replied.

Trey and I knew very well what would happen when I signed the death certificate. Brain death is a firm, unambiguous, and operationally solid determination, an absolute point of no return for the brain. Any two competent neurologists or neurosurgeons who examine a brain-dead patient will come to the same conclusion, just as we had: this entire brain will never recover, and all the king's horses and men can't do a damn thing about it.

The problem is the word *dead*. It muddies the important issue, as does *diagnosis*. Brain death is not a diagnosis—a word that suggests probability—but rather a *determination*. A *diagnosis* raises the specter of false positives, of fallibility, of someone being buried alive. That can only happen if someone does the test incorrectly, and we hadn't.

"Look, Trey," I said, "it's fine to have an operational definition to work with. We couldn't get through the day without that. But you are in a position, because you are a doctor of the brain, to think about these things more broadly, and you should, because if you don't, nobody will."

By then it was after dark. We were standing in the ICU outside of Mike Kavanagh's room. For an uncomfortably long time, Trey said nothing, did nothing, looked at nothing. Then he made an abrupt about-face, and walked out of the room. It wasn't a yes, it wasn't a no, it wasn't dismissive. But he left a vacuum behind, and it was quickly filled by Elliott, who just happened to be walking by.

"Is he dead yet? The nurses want to know. They want him out of here. The transplant surgeon is lurking."

"I'm signing the papers now."

"Well, then," said Elliott, without irony, satisfaction, or relief, "bring out your dead."

In *The Wizard of Oz*, after the tragedy of the tornado and the falling house, the medical examiner of Munchkinland was called upon to check under the front porch for the remains of Evanora, the Wicked

Witch of the East. After due consideration, he solemnly pronounced
to the mayor:

> *As Coroner I must aver,*
> *I thoroughly examined her,*
> *And she's not only merely dead,*
> *She's really most sincerely dead.*

In the case of Mike Kavanagh, Trey was satisfied with "merely dead,"
as was the Presidential Commission, the Commonwealth of Massachusetts, and the Vatican. I'm not entirely sure.

13

Boats Against the Current

Based on a true story

"Gilbert thinks he wants to be an ophthalmologist."

Elliott says this to the window plant as we drift in an eddy of late-afternoon, end-of-the-week inertia. It is Friday, and Elliott has tickets for the symphony. I have a dinner party to go to. We are in my office, kicking off the weekly ritual that will, should the gods find favor in it, restore our enthusiasm for small pleasures.

Gilbert is one of the third-year medical students who have been rounding with us. He's a short fellow, wears circular, wire-rimmed glasses of the 1920s style, and may or may not have a slight British accent. It's hard to tell, he speaks so softly. Yet he has a way with patients, Elliott and I have noticed, a way that is difficult to quantify and impossible to teach. It is, in part, a delicate balance between listening, processing, appearing to care, actually caring, and giving the patient what he or she needs, even if it may not be what they want. It is not that he speaks their language, but that he has the knack for translating what they tell him into usable form, then explaining it back to them in a reassuring way—a rare, seemingly innate skill. It doesn't hurt that he looks like Harry Potter.

"I saw this guy today with an impulse control disorder," I tell Elliott, "a Parkinson's guy. Very interesting. He was going around punching walls, tailgating people because they cut him off. He wanted to get out of the car and sock somebody." That's how a patient frames a problem in his own vernacular.

"Too much dopamine agonist," Elliott says, reframing it in ours.

"Correct. Too much Mirapex, so I lowered his dose."

That's what makes medicine medicine. People are letting us in, but we can't expect them to describe their problem in medical terms, so we reframe what they say in a way that's usable. We give it coherence. If somebody comes in with numbness or paralysis, with headaches or speech problems, with tremor or confusion, we take the patient's report, bring it to the level of brain and nervous system operations, then come up with a plan. It can be a very intense chess game—speed chess—because on rounds, we have to act fast. The value we add is the intellectual act of making those moves, and the clarity with which we can do it.

Intuitively, Gilbert seems to comprehend this, and has the potential to put it into action, and yet he has rounded with us long enough that he also sees the price he would pay if he chose neurology over ophthalmology. Neither Elliott nor I need comment on what Gilbert's pseudo-decision means: that ophthalmology would provide him with a comfortable living, a well-rewarded, relatively stress-free career compared to neurology. He is facing a big decision, yet we understand. Gilbert has glimpsed the uphill road ahead, and has overheard enough conversations like the one Elliott and I are having now.

"That woman I saw last month," Elliott says, "almost ninety, big stroke, then she got pneumonia. Then we had the family meeting. Flavio was with me. We said, 'She's eighty-nine, and at best, if she recovered, she'd be completely paralyzed on her left side, and completely dependent for everything.' There were two sons who were very outspoken, telling us that Mom didn't want to live this way. They said, 'Please don't do all these things to her.' They want comfort measures only. Everyone is in agreement except the daughter, this quiet, spinsterish type, a bit of an oddball, sitting in the corner. She says nothing. The social worker is in the room, a junior resident is in the room, Flavio is in the room. The woman dies a few days later. Two weeks after that a complaint is made against me and Flavio, that we unnecessarily caused her death, that the daughter didn't understand how her mother could be getting better, then all of a sudden could get so much worse that she died. She says we didn't give her a chance. She's really angry. Only now does she speak up to say that we didn't give her mother a chance."

There's nothing to say except what Elliott already knows: that nothing will come of it except stress and aggravation.

"And how was your day?" he asks me.

"I saw a woman who's going to end up being a disaster. She's got a lung mass, a cyst in her bone that looks like osteosclerotic myeloma, and a six-month progressive course of ataxia and spasticity. I don't even know where to start. The hematologists are there, the orthopedists are there, the pulmonary guys are in there. She was supposed to see me at the office next week. The husband called me, frantic, and told me, 'She can't walk!' So I said, 'Bring her to the emergency room,' and they admitted her, which was the right thing to do. Then I heard from Elena that Gordon Steever, the bowling alley manager, died last week up in Saugus. And then, get this, I happened to be walking through the ward, and there was this psychotic woman, a Trinidadian woman up there in her little johnny, and I was the only one who could calm her down and lead her back into her room. She was about to scratch someone's eyes out, and I calmly put my hands on her shoulders and steered her like a car, and led her back into the room. Then a social worker comes up to me and says, 'You know, I don't think you ought to be touching her like that.'"

"And so we beat on," says Elliott, "boats against the current."

. . .

Arwen Cleary, the ice skater with multiple strokes, returned home and lives a cloistered life with the help of her daughters and her boyfriend. Her right visual field is permanently impaired, yet her occipital cortex does not register that fact, even though her frontal cortex does. She did not get the surgery to address the growth on her heart valve; the growth seems to have disappeared. Why? We don't know. We have to hope that the blood thinners will prevent any further strokes. I have suggested that she allow us to fit her with glasses that will reveal the right side of her world to the functioning left side of her visual field, and she has yet to take me up on it.

Mrs. Gyftopoulos has returned to a fairly normal life, as have Vincent Talma, and Cindy Song, although they all have some deficits: a hole in the head, a barely detectable speech problem, a missing ovary. During the time they passed through the wards, most of their fellow patients came out okay: the strokes, accounting for about 20 percent of our cases, usually resolve, and future ones are staved off with blood thinners. The aneurysms, most of them anyway, are neutralized through surgery. The seizures are treated with various antiepileptic drugs. We cleared the patients off our lists one by one. Some of them will have follow-up appointments as outpatients, many of them we will never hear from again. The only cases whose long-term resolutions I can be entirely sure of are those that ended in death, and even then, as in the case of Gordon Steever, the candlepin bowling man, I don't always find out how it happened.

It is at this point that I should confess that there is no candlepin bowling alley in Dorchester, that Vincent Talma was not playing softball when he became aphasic, that Louise Nagle did not go to Cornell, that Ruby Antoine was a laborer, but not a mechanic, that Tikvah does not live in Yorkville, that that is not her real name, that all of the other patient names, save one, along with the names of local community hospitals, were invented. We have tried to tell these stories truthfully in all essential elements, if obliquely in the inessentials, which is to say that for the sake of confidentiality, the patients we have described, with the exception of Michael J. Fox, are mere analogues of actual people, yet neurological doppelgangers. The residents and many of the other physicians named and described here are also avatars, but the situations are true enough, the diseases, disorders, lesions, and tumors are painfully real, the dialogue is 70 percent verbatim, 20 percent recollected, and 10 percent extrapolated.

This is also where I should point out that this is not an "as-told-to" book, but rather a co-written, co-experienced work. It had to be. It required two sets of eyes, two entirely different perspectives. I would not have been able to tell these stories on my own because, quite simply, I was too focused on helping each patient to see beyond their immediate medical problems. No physician, in the moment, can simultaneously serve the patient and the story. This book was made possible by someone who was willing to experience life on the ward with me, even without me, and merge our combined experiences into one voice, someone who kept notes, someone who listened to the patients and residents as closely as I did, and heard what I did not, someone who bothered to look out of the window, someone who was curious enough to keep asking, "What was that about?" and "Whatever happened to . . . ?"

Gordon Steever did die in a psychiatric nursing home four months after being discharged from the Brigham. His brain was not autopsied. We still do not know what caused his confusion, and we never will. Wally Maskart went home after a two-month stay at a psychiatric facility, where he showed steady improvement. He continues to care for his wife and manage the family finances. He remembers little of his stay at the Brigham, except that his thinking was fuzzy. He no longer thinks I am Sanjay Sanjanista, but he still insists that he was at Game 6, and saw Dwight Evans make The Catch, and I believe him. It turns out that he did not sell, but rather kept, the locomotive—the vintage Lionel Red Comet that he bought in a pique of mania—and he has no regrets about the expense. Old Doc Vandermeer died from complications following the surgery to remove his meningioma. It was worth trying, because the growing tumor would have further clouded his mind. Michael J. Fox went on to star in a new television series in which he plays a news anchorman who is diagnosed with Parkinson's, and has to reorganize his life in order to deal with that fact. It is a comedy. The idea, he told us, is "to put it out there and say, 'This is my life. This is what it's like to live with it.'" At this stage in his career, he adds, "I can do anything, I can play anybody, as long as they have Parkinson's."

"The pope?"

"I could play the pope."

Ultimately, he did give me a Ferrari, a Testarossa, not made in Italy but rather made in Japan by Mattel. "But vintage!" he said, by way of consolation.

In the 1990s, when I saw my mentor, Raymond Adams, failing physically due to congestive heart failure, I saw what he represented fading like a painting in the sun, and I wondered whether I should try to sustain his exquisite clinical-pathological correlation method through a dark age, on the chance that there would be a renaissance, and that it would come back into favor. For a time I thought, *Yes, that is what I will do.* Then I woke up and went my own way.

Raymond Adams, C. Miller Fisher, and even Doc Vandermeer belonged to a generation of physicians who were so wrapped up in the ethos of what they were doing that they inadvertently gave rise to cults of personality that warped the sensibilities of their acolytes. The ethos served a social good, so the cults survived for time, but have now made way for a collective caring for patients in which process has supplanted personality. Part of my role today is regrettably to lead the next generation of neurologists through the transition from the old school to a new one, to turn them into interchangeable parts in the new health-care system. Yet I want them all to be best in show. I want them to live the *vita medicalis*, and get their neurology from their patients' stories, not from books and Web sites. Otherwise, they will be merely part of a dystopian and impoverished version of medicine that is disconnected from life, from suffering, even from death.

I am glad that Hannah got to know Doc Vandermeer, if only in passing. Like my mentors, he was one of the greats of twentieth-century medicine, and he saved countless lives by dedicating his own life, sacrificing it in a way, to the relief of suffering. She got to see the mannerisms of the Old Guard, for better and worse: the insufferable circumlocution, the lifestyle bordering on self-negligence, the elitism tucked behind a mask of benevolence, but also a dedication to science and service verging on the messianic. It also gave her some perspective on me.

I am now in my sixties, a fact that weighed on my mind when I first met Doc Vandermeer on the ward. Our encounter stirred up thoughts, many of them tough to face, of obsolescence, of becoming an anachronism. I can take some solace in knowing that the careful, thoughtful, involved, committed clinician, the one dedicated to service and the art of neurology, will survive in people like Hannah, and will be carried on with a completely different style than my own, refreshingly so, and in many different places.

Hannah had hoped to stay on at the Brigham. Instead, she took a position as a fellow in neuro-infectious diseases at another top program, far from Boston. She was disappointed about leaving, but I thought it was for the best. Wherever she goes, I want her to be the person whose opinion is sought out above all others, the one who is told: "Dr. Ross, we have a case of severe headaches, drop attacks, and seizures." And instead of saying, "Oh, the differential diagnosis is such-and-such," she will say, "What are the seizures like? Tell me about the seizures." She will query the patient, and she will show the residents that if you are willing to take the time to ask and to listen, the patient will usually have the answers.

I ran into her recently at a medical conference, and she told me how, as a junior faculty member, she now has to conduct Chief's Rounds when her department chairman is away. "Whenever the residents present a case with fever and back pain," she said, "I will say to them, 'What is the one thing that will kill this patient if you miss it? What is the one thing you want to make sure it is NOT?' They don't know what to say because they're supposed to be tricking me. They presented a patient with West Nile virus, and when I went off about protecting the downside, they looked at me like I had ten heads. But Harry Connaway's autopsy changed everything for me. Ever since that autopsy, I see that ocean of pus in my dreams, and I say, 'Don't think about epidural abscess the way you were taught, because the way you are taught is not the way it happens. I know. I've seen it up close and personal.'"

The cases described in this book, with the obvious exception of the reminiscences and flashbacks, occurred within a span that covered two services on the neurology ward and two on the ICU. While each case was unique, the services themselves were representative. Right now, for example, as Elliott and I sit talking, we could go upstairs and find a comparable assortment of cases on our wards: some mundane, a few unprecedented, and at least one utterly fascinating. And they just keep coming.

Three men have been admitted, all in the same week, with remarkably similar symptoms. The first is a college student who was studying architecture abroad and began to have painful numbness in his feet and hands, with trouble walking. In the previous three weeks, his hair had begun to fall out. The second is a forty-one-year-old man who was admitted with an unusual blistering rash and confusion. He began to complain of a headache, and threw up repeatedly. The third is an Indian engineer who had two seizures at a pizza parlor, and came in very confused. All the tests were essentially normal, no one in the Emergency Department could sort it out, so they sent him upstairs to us.

As it turns out, all three men were poisoned over a period of weeks: the student purposely with rat poison containing thallium by a fellow student with strong anti-American views; the second with mercury by his wife; and the engineer by an Ayurvedic skin cream, his chosen treatment for psoriasis, that just happened to contain arsenic. The moral of the story, I tell the residents, is that people are being poisoned by others and poisoning themselves all the time. Check for heavy metals, even in your sister.

Last Wednesday, a twenty-four-year-old plumber's assistant came in. He said, "This is terrible, Doc. My arms have been getting weak, and now both of my hands are getting weak." He had a dozen MRI scans at other hospitals, all of them normal. Again, nobody had a clue. They told him he might have ALS. His mother was frantic.

I did the exam, and said, "You've got a problem called Hirayama disease. Not seen much in the United States. Mostly in Japan. The ligament on the back of your spinal cord is buckling when you bend your neck forward, and all of your scans were done with the neck straight."

I sent him for another MRI, and they called me from the X-ray department to say, "You're wasting our time, it's normal."

I said, "Did you flex his neck?"

They said, "We don't do that in the MRI scanner. That's not the protocol."

"You don't get it," I said. "The disease is defined that way."

I bring the new scans up on my monitor to show Elliott. This is one of those cases where the right picture really is worth a thousand words, and the wrong picture is worth one. I show Elliott what the MRI looked like when they had him flex his neck. The ligament buckles, it bunches, it causes venous congestion, and it pinches the spinal cord. "Every time the kid leaned forward he compressed his spinal cord, and he's a plumber. What it's done over a year or two is damage the anterior horn cells in the cord, and caused what looks like ALS, but most definitely is not."

There are eight million stories in the naked city, and we get our fair share of them. The narratives flow forth every morning when the team convenes in the conference room:

"Miss Staines is our seventy-four-year-old lady with COPD and a stroke. She's going to need a speech-and-swallow consult this morning."

"Mrs. Henson is an eighty-eight-year-old lady with history of colorectal cancer, who also had a left MCA stroke in a setting of A fib. She had an episode last night where she was a little confused."

"Dorothy Fitch has tingling in her toes after a GI illness, but normal reflexes. We still think she has Guillain-Barré syndrome."

"Miss Tannenbaum, the twenty-nine-year-old lady with MS coming in with optic neuritis, blind in her right eye, she's on p.o. steroids. We were going to talk to the outpatient MS doctor."

"Kerry Norris, age nineteen. Complex partial seizure generalized into a tonic-clonic seizure. They gave her some Ativan and restarted her antiepileptic meds."

"Eric Servi, thirty-eight-year-old man with congenital heart disease. He had a very strange anatomy of his heart as a child and has had many surgeries to fix it. He also had a stenotic subclavian. He came in with dizziness, and had a tiny punctate cerebellar infarct from an acute versus chronically thrombosed left vert. Wasn't comfortable leaving so he stayed overnight."

"Mr. Comstock is a sixty-five-year-old man with a history of squamous cell cancer under his left eye, likely an infiltrative tumor in the left orbit. He has been seen by just about everybody. No one wants to operate. We'll need to go up the chain to the neurosurgeons."

Welcome to the Ellis Island of nervous diseases, where the tired, the poor, the embolic, the metastatic, come through in wave after wave, and we examine them on every level, not just with scans or tests. In this brave new world, our approach has to be integrative, it has to be synthetic, it has to be elegant. It may seem messy at times, but neurology is fascinating partly because of its messiness, because it imposes order on chaos. And it is still the Queen.

"I got a call the other day during dinner," I tell Elliott. "It was Callie, and she said, 'I have this patient who has been in and out of the hospital with seizures. He's a complete wreck, unmanageable, and his tumor is end-stage and there is a ton of radiation necrosis.' I said, 'Sure, admit him.' We put him on steroids, and he becomes calm, an almost serene bodhisattva."

The fact that nothing bothers him suggests that his agitation was caused by the radiation, not the tumor. He is not at the level of hospice care. Steroids will keep him in line. He will die in a year or two if it's radiation damage, and within months if it's a tumor, but either way, his brain is altered enough by the disease or the radiation that he cannot have a true emotional reaction to what's happening to him. He is in no position to appreciate on a visceral level the gravity of his condition. I could ask him, "Do helicopters eat their young?" And he would say, "Yeah, sure . . . probably . . . whatever." I could tell him that he is going to die in a week, and he would say, 'Oh, really! Die in a week? Okay.' "

That might sound philosophical, but it is not. It is the response of a damaged nervous system, and in his case, damage to the place that supplies him with an awareness of threats to his existence.

"He lacks insight," I tell Elliot, "and even seems to know that he lacks insight."

"How is that possible?" Elliott says. "How can you be aware that you don't have insight?"

"I'll have to get back to you on that, but apparently you can."

People walking down the sidewalk in front of the hospital might find that question interesting in theory, but the minute they pass through the revolving door, all they care about is one thing: "Am I going to live or am I going to die? You better figure it out for me. I don't care how you figure it out. There may be eight million stories in the naked city, but I only care about one of them. I care about my story. How's my story going to end?"

What I would say to them is, "I hear you, I'm working for you, working to make you better now, working to help you survive, even if the news is bad." What I would like to add is, *"and you can't imagine*

what goes into that." It's not like being a butcher, a baker, or a candle-
stick maker. We're trying to understand how the brain works so that we
can fix it. We do the job, but we don't dwell on high-flown philosophical
implications: What is consciousness? Where is the soul?

We may be interested in such questions, but up on the ward we
don't care because every grand theory of mind and every sweeping
generalization about consciousness falls apart when exposed to the
cold, hard facts of a single case. Instead of theories, we need clinical
truths: What's wrong with this person's brain? Can we get it back to
normal or some semblance of normal? Can we at least get them on the
right track? When we can't, the question turns grim: How will I walk
this patient and this family through this all the way to dementia or
death? Because the job doesn't end with the diagnosis.

Once a week, we make the trek up to the neuropathology lab, where
the residents each take one of the eight places at the teaching microscope while the Chief of Neuropathology runs through the histology
and morphology of biopsied tissue from some of our tumor cases. The
stained tissue samples hold the evidence. The jury could go either way.
Elliott usually gives me a meaningful look when bad news seems imminent. Then I know he is thinking of Max von Sydow in *The Seventh
Seal.*

We all invent notions to reassure us and lies to protect us. Elliott
calls these distancing techniques. His are largely cinematic. "We
have just spoken to Death," he sees himself telling the patient, "and
Death has told us your time has come."

Death, for Elliott, is the Chief of Neuropathology. What we hear
Death say up in the lab may sound opaque to the man on the street,
but not to us.

"Notice how very densely cellular the tumor is," the Chief intones.
"It displays pleomorphism of nuclei, which means nuclei of different
sizes and shapes, and then mitoses, very many, and thickened blood
vessels with prominent endothelial cells."

The Chief is a refined and cultured man, descended from European royalty, who is oblivious to the fact that Elliott has cast him in a Bergman film. He delivers his lines in a detached and didactic way. "We haven't found necrosis, but already this meets the criteria for glioblastoma. You can see the tremendous crowding, the piling up of cells, a very serious problem. The gradation of astrocytomas is one, two, three, and four, and for four, the highest grade, it includes the presence of mitoses and cellular proliferation, which we have in this case."

All very interesting, all very edifying and matter-of-fact, but at the other end of the hospital, Mr. Gerrity, a retired firefighter with a close-knit family, the man whose brain yielded up this specimen, awaits the verdict, and Death has delegated to us the task of breaking the news. For the Chief, it's a wrap, his job is done. But for Mr. Gerrity, for his family, and for us, it's Pearl Harbor, and World War II is just beginning.

I have a friend who is one of the leading trial lawyers in Boston, a specialist at handling murders, and he was very down recently because he had just lost a case. Even though it was clear to me and to him that his client actually did the crime, he felt bad because the guy was going away for life.

"I can sympathize," I told him, "because I know what it's like to lose one. It's an irrevocable life sentence, and you will always wonder if you did enough."

When I go to the ICU, I routinely encounter situations in which a slipup can deal somebody a life sentence, either being crippled or dead. During my ward service, chances to slip up might occur several times in a day, and because I have to work fast, my exposure to this kind of downside loss is enormous. I have to go without the contemplative time afforded a trial lawyer: time to scrutinize and adjudicate every detail, time to follow clearly enumerated rules of evidence, time to adjourn if necessary. Every day I might go through what my lawyer friend goes through three or four times a year, with the same kind of outcome, on a knife's edge, and at the same level of responsibility.

What is it like to be always on the brink of your next big mistake? How do you manage as a physician in a high-stakes game in which you are definitely going to make some people worse? All you can hope for is a very much higher proportion of people you make better. You can't be in this business if you can't live with the risk and the deep disappointment of bad outcomes.

How would I sell *that* to Gilbert? I wouldn't even try to sell it to him. He is in medical school. He will figure it out. Instead, I periodically have to sell it to myself by fighting off the cynical realization that Disease (with a capital D) almost always wins, that we only occasionally win. We insist that this is not the case in order to maintain our sanity, if not our molecular structure. But that's all right. A day at the hospital is not transformative. At best, I can walk out knowing that relief of suffering is what we're good at, even if society needs to believe that what we're good at is cure. It is to everyone's benefit that society believes that. All of us—patients and doctors—cultivate the fiction that science conquers all, that it can provide the cures. How could God have created a world in which it can't? We need to sustain our faith in science, our paradoxical belief in its divine power. We have always had to believe that, going back to pre-Hippocratic times. "I'm going to give you a potion, it will solve your problem." That statement promises some control over mortality and destiny. That's why what we do is as much shamanistic as medical, because society cannot relinquish that hope, the belief in the curative power of something, of medicine, of prayer, of diet, of therapy, of sheer expertise, of connecting with another human being, rather than acknowledge that the universe is like the eye of a dead fish: cold, uncaring, unreflective, unresponsive.

"Have you ever noticed," I ask Elliott, "that during a code everybody is huddled around, everybody has a job, everybody's engaged, and the focus is on both the patient and the electrocardiogram? But the instant it's over, the moment they call a failed code, everybody just turns on their heels, and if you're still there, spending a last minute with the body, you see this amazing thing, almost like a ballet. They just drop everything and walk away, and you see their backs."

"Disengagement."

"Exactly. They're gone, yet it's discomfiting. I started hanging around after codes because I've been interested in the size of people's pupils when they've just died, and I get to see this little ballet."

"You shouldn't mess with death," Elliott said.

"That's what Trey tells me. I tried to lure him into a discussion about brain death, but he doesn't see the point."

"What *is* the point? Brain death is death. End of story."

"But people don't get it."

"Well," said Elliott, "knock yourself out. The rest of us have work to do. I'll wait for it to come out on DVD."

Elliott is a brilliant diagnostician and a keen judge of character, but a detached observer. He doesn't try too hard to connect with patients, with his own colleagues, or even with me. He is an expert at "speed rounds," for which the residents love him because he can get them in and out of a ward with twenty patients in about forty minutes. He is adored by the nurses merely for gracing the place with his presence, as though he has taken time out from a busy schedule just to say hello, pay a compliment, ask about the new baby, notice a pair of shoes, enter a betting pool, or criticize the Red Sox management. This is how he handles the pressure. He dresses well, he cultivates fine habits, and he lives far away from this madness. He is Jay Gatsby, without the insecurity complex. Beyond that, I can't say that I really know who he is.

Elliott is my friend in the way that several of my colleagues are friends, even if he has a bad habit of showing up at the wrong time in order to tell me something I could live without knowing: that my patient is a child molester, for example. He is nothing like me, but I would rather have him around than not. I don't need to be his closest friend, but I do trust him entirely. He is a very good doctor. Almost nothing gets by him. Judged by any of the currently fashionable measures of outcome, he would score extremely high. He has a tremendous command of therapeutics. But there are patients who do not want to go back to him.

"He didn't listen," some of them say. "I told him I was having this trouble with my arm, he told me to get up on the table, he said this is what you have, this is what you need to do, and that was it."

What they mean is, he doesn't hear them out, he doesn't savor their stories. That's his style. Mine is to pull up a chair, put my hands behind my head, lean back, and just listen, as if to a book on tape. Almost never are two stories the same: patient X and patient Y may have the same medical problem, but very different personal experiences.

What Elliott cannot abide about neurology and about the hospital is the transactional nature of the job, the lack of closure. He moves from one case to the next without getting emotionally caught up in any one of them. Ironically, Elliott never tires of hearing *my* patients' stories from me. He always has to know the rest of the story, and when he can't find out what happened from my end-of-the-week recaps, he tries to imagine how the story ends. It is part of his cinematic sensibility. He should have been a screenwriter.

"What happened to Gordon Steever?" he asks me.

"He's dead," I say, hoping to head him off, but he won't have it.

"I know, but how?"

"No idea."

"Did we ever find out what he had?"

"No," I say. "No one claimed the body, no one asked for permission for an autopsy."

"We know what bowling alley he used to manage, though, right?"

"Yes, we do."

"What about going down there and asking around about him."

"That would be against the privacy rules," I tell him, but I know what he's building up to.

"I don't see how bowling could be a violation of anybody's privacy. Let's go. How about tomorrow? Two strings. You roll on Shabbos, don't you?"

Elliott has no intention of going bowling. He will probably spend the day reading, cooking, drinking Scotch, and periodically looking out over the Atlantic Ocean from his arts-and-crafts home in Manchester-by-the-Sea. But I know that he's been waiting all week to ask me that question, maybe all month. And as he knows very well, I roll on any day of the week.

ACKNOWLEDGMENTS

The extraordinary way in which the disordered human brain spins its tale of woe reveals how the organ works, how it creates the primal human experience of illness. Over the years I have been privileged to listen to countless such narratives, to translate them not just into syndromes (cerebral hemorrhage, multiple sclerosis, motor neuron disease), but into truer categorizations of a patient's plight (confused mind, difficult family, impossible case, threatening crisis, irreconcilable opinion, slow death, tale of heroism). My patients are compelled to get their stories out to assure themselves that there is a steady hand attending to their deepest concerns, and this book is my way of thanking them for being so open. I appreciate the generosity of my colorful mentor, Dr. J. P. Mohr, and of my joyful colleague, Dr. Dan Longo, who allowed me to incorporate snippets of their own patient stories into those of my own.

My residents-in-training are the main reason I get up early to come to the hospital. Several of them allowed me to take liberties with our time on the ward together in order to accumulate the material for this book. These residents are precious commodities. Foremost among them are several who participated during the writing of the book, but who I will not identify in order to preserve their privacy, but I want them to know how much I appreciate their generosity.

Having taken notes during many patient encounters, and encouraged by my coauthor to record many others, I saw this book as a chance to retell compelling patient narratives through the knowledge of neurological diseases. Readers of the manuscript wanted to hear more about how I cope with difficult cases, what was I thinking at various junctures, and what the practice of neurology means to me. I respond by declaring that my motivation and satisfaction lie in savoring each patient's story, in listening to the brain itself. Those readers were friends, family, and colleagues who were kind enough to read the book at various stages. We found a considerable amount out about each other in our conversations about the manuscript, and I offer my sincere thanks to them: Jennifer Lyons, MD, Eileen Bockoff, uncle Harry Schachter, David Fine, Alan Schlesinger, Roger Cassin, Joseph B. Martin, MD, Chuck and Rivka Raffel, Kathryn Giblin, MD, Martha Neagu, MD, Thomas Moorecroft Walsh III, MD, Martin A. Samuels, MD, Dr. Geoff Greif, Dolores Araujo, my most capable assistant, and Barbara and Michael Lissner. But the person I drove crazy when writing the book was my wife, so hail to Sandy.

I met Brian Burrell as a result of reading his wonderful book, *Postcards from the Brain Museum*. I invited him to discuss the book at one of the book clubs that our department organizes for the neurology and psychiatry residents. Brian and I have similar outlooks, and he has a lyrical talent for making true stories sing. Whenever I drifted toward standing on a soapbox, and ranted about the everyday frustrations of medicine, he brought me nicely back to the patient's words in a way that enlivened and enriched my practice, for which I owe him a debt.

We are both grateful to our respective institutions, the University of Massachusetts at Amherst and Brigham and Women's Hospital,

which afforded us the freedom to step outside our usual roles. We also wish to thank Michael J. Fox and Dwight Evans for permitting us to use their stories. Finally, we should acknowledge the role of serendipity, which allowed us to relate these stories more or less as they occurred.